80 Years of Zentralblatt MATH

Olaf Teschke · Bernd Wegner · Dirk Werner

Editors

80 Years of Zentralblatt MATH

80 Footprints of Distinguished Mathematicians in Zentralblatt

Editors
Olaf Teschke
Zentralblatt MATH / FIZ Karlsruhe
Berlin, Germany

Prof. Dr. Bernd Wegner
Technische Universität Berlin
Fachbereich Mathematik
Berlin, Germany

Prof. Dr. Dirk Werner
Freie Universität Berlin
Fachbereich Mathematik und Informatik
Berlin, Germany

ISBN 978-3-642-21171-3   e-ISBN 978-3-642-21172-0
DOI 10.1007/978-3-642-21172-0
Springer Heidelberg Dordrecht London New York

Library of Congress Control Number: 2011937025

Mathematics Subject Classification (2010): 01-06, 01A55, 01A60, 01A61, 01A65, 00A17, 01A74

Printed on acid-free paper

Springer is part of Springer Science+Business Media (www.springer.com)

Preface

Founded in 1931 by Otto Neugebauer as the printed documentation service *Zentralblatt für Mathematik und ihre Grenzgebiete*, Zentralblatt celebrates its 80th anniversary in 2011. Today it is, under its new name ZBMATH, the most comprehensive reference database in mathematics, and many famous mathematicians have been involved in this service as reviewers or editors. Zentralblatt has covered their work as a part of its general documentation activities, dealing with all mathematicians worldwide. All mathematicians have left their footprints in Zentralblatt in a long list of entries describing their research publications in mathematics.

On the occasion of Zentralblatt's 80th anniversary we are presenting in this book a selection of reviews, one from each year. Many of them are connected with a prominent mathematician related to Zentralblatt as a member of the editorial board or as a reviewer; names like Courant, Kolmogorov, Hardy, Dieudonné, Hirzebruch, Faltings and many others can be found here. In addition to the original reviews, the book offers the authors' profiles as generated by the ZBMATH database, indicating their co-authors, their main research areas, their favourite journals, and the time span of their publication activities. Furthermore, a generously illustrated essay by Silke Göbel describes the history of Zentralblatt.

Berlin, June 2011

Olaf Teschke
Bernd Wegner
Dirk Werner

Glimpses into the history of Zentralblatt MATH

Silke Göbel

Formation and early successful years

In 1931, the first volume of *Zentralblatt für Mathematik und ihre Grenzgebiete* was published. It presented the bibliographic data of all recently published mathematical articles and books, together with peer reviews by mathematicians from all over the world. Zentralblatt became the second comprehensive review journal for mathematics in Germany—the *Jahrbuch über die Fortschritte der Mathematik*, established in 1868, was the first reviewing journal in mathematics worldwide [MO68].

In the *Geleitwort* (preface) to the first volume, the intentions of Zentralblatt are formulated as follows [Neu31]:

> Das „Zentralblatt für Mathematik und ihre Grenzgebiete" stellt sich die Aufgabe, in rascher und zuverlässiger Weise über die gesamte Weltliteratur der Mathematik und ihrer Grenzgebiete in zunächst monatlich erscheinenden Heften Bericht zu erstatten. Wie schon der Titel sagt, bilden die mathematischen Wissenschaften den Schwerpunkt des neuen Referate-Organs. Trotzdem sollen aber die „Grenzgebiete" durchaus ebenso ernst genommen werden wie die sogenannte reine Mathematik.[1]

Although Zentralblatt had essentially the same agenda as the Jahrbuch, there were differences. Zentralblatt was published several times per year. An issue was published as soon as sufficiently many reviews were available,

---

1. "Zentralblatt für Mathematik und ihre Grenzgebiete" aims to publish—in an efficient and reliable manner—reviews of the entire world literature in mathematics and related areas in issues initially appearing monthly. As the name suggests, the main focus of the journal is mathematics. However, those areas that are closely related to mathematics will be treated as seriously as the so-called pure mathematics.

O. Teschke et al. (eds.), *80 Years of Zentralblatt MATH*,
DOI 10.1007/978-3-642-21172-0_1, © Springer-Verlag Berlin Heidelberg 2011

resulting in a frequency of three or four weeks. Zentralblatt was shorter and more objective, and reviews could be written in English, French or Italian as well as in German. The Jahrbuch first collected the reviews and sorted them in subject chapters before publishing the entire annual collection. While Jahrbuch maintained the principles *completeness* and *classification of all articles of one year*, Zentralblatt counted on *promptness* and *internationality*. As a result of World War I and the amount of mathematical literature that grew explosively during the 1920s, the principles of the Jahrbuch led to a big backlog in their editorial work. Mathematicians had to wait for a long time—four years on average—until they could read about the latest literature.

From left to right: *Otto Neugebauer*, *Richard Courant*, *Harald Bohr* and *Ferdinand Springer*.

Mathematicians Richard Courant, Otto Neugebauer and Harald Bohr, together with the publisher Ferdinand Springer, took the initiative for the foundation of a new mathematical reviewing journal. Harald Bohr, the brother of the famous physicist Niels Bohr, worked in Copenhagen. Courant and Neugebauer were professors at the University of Göttingen. At that time, Göttingen was considered one of the central places for mathematical research, having appointed mathematicians like David Hilbert, Hermann Minkowski, Carl Runge and Felix Klein, the great organiser of mathematics and physics in Göttingen. His dream of a building for an independent mathematical institute with a spacious and rich reference library was realised four years after his death. The credit for this achievement is particularly due to Richard Courant, who convinced the Rockefeller Foundation to donate a large amount of money for the construction. The contacts between Courant and the Rockefeller Foundation were, in turn, established with the help of Harald and Niels Bohr [Sch83].

Courant, Neugebauer, Bohr and Springer knew each other for many years; Richard Courant, for instance, met Springer in 1917 and later worked as an advisor to the Springer publishing house. On many occasions like conferences, private meetings and walking tours, they thought about improving the working environment and the information and communication infrastructure for the mathematical community.

The *"Mathematisches Institut"* in Göttingen, built in 1929.

Ferdinand Springer had already rendered outstanding services to mathematics as one of the main publishers of mathematical books and journals, including *Mathematische Annalen* and *Mathematische Zeitschrift*, which to this very day are among the best international peer-reviewed journals in mathematics. In 1923, their editors addressed a letter of thanks [Sar92] to Springer saying

> Wenn nicht Ihre opferbereite Unternehmungslust und Ihre umsichtige Energie sich der Sache der mathematischen Wissenschaft angenommen hätte, so würde heute die Mathematik in Deutschland nicht mehr lebensfähig sein.[2]

Among the addressees were Felix Klein, David Hilbert, Richard Courant, Max Born, Ludwig Bieberbach, Albert Einstein, Edmund Landau and Arnold Sommerfeld. In 1930, the Faculty of Mathematics and Science of the University of Göttingen conferred an honorary doctor's degree on Ferdinand Springer.

Zentralblatt's first editorial board consisted of Pavel S. Alexandrov (Moscow), Julius Bartels (Eberswalde), Wilhelm Blaschke (Hamburg), Richard Courant (Göttingen), Hans Hahn (Vienna), Godfrey H. Hardy (at that time Oxford, later Cambridge), Friedrich Hund (Leipzig), Gaston M. Julia (Versailles), Oliver Kellogg (Cambridge, MA), Hans Kienle (Göttingen), Tullio Levi-Civita (Rome), Rolf H. Nevanlinna (Helsinki), Hans Thierring (Vienna) and Bartel L. van der Waerden (Leipzig). Otto Neugebauer became the first editor-in-chief. The first editorial office of Zentralblatt was on the premises of the Springer publishing house in Linkstraße in Berlin.

The editors and authors of the publications in Zentralblatt were not the only well-known mathematicians associated with Zentralblatt. Distinguished researchers from various countries belonged to the large group of reviewers,

---

2. If you had not attended to this science with such dedicated enterprise and prudent energy, mathematics would not be viable in Germany today.

including Enrico Bompiani, Charles Ehresmann, Hans Freudenthal, Otto Haupt, Helmut Hasse, Heinz Hopf, Friedrich Hirzebruch, Edmund Hlawka, Andrey N. Kolmogorov, Đuro Kurepa, Emmy Noether, Johann Radon, Erhard Schmidt, Francesco Severi, Wacław F. Sierpiński, Teiji Takagi, Béla Szőkefalvi-Nagy and Edward M. Wright.

Unfortunately, a large amount of the editorial correspondence documenting the foundational time of Zentralblatt got lost in the turmoil of the Second World War and the post-war period. Neugebauer burnt most of the editorial correspondence when he emigrated to the USA; the only records of the journal he managed to save were the cumulative indices. Nowadays, some letters from those times are kept in the archives in Berlin, Göttingen, Heidelberg, New York and Washington; three folders documenting the time between 1947 and 1963 can be found in Zentralblatt's current editorial office. Within very few years, Zentralblatt became a successful journal with 18 volumes appearing between 1931 and 1938.

The years 1933–1945

In 1933, shortly after the Nazis assumed power, Zentralblatt experienced a critical change. On April 7th, the *Gesetz zur Wiederherstellung des Berufsbeamtentums* (the Civil Service Restoration Act) was enacted which banned Jews and political enemies from holding jobs as civil servants (§3, §4). Three classes of civil servants were excluded from the ban, among them the veterans of the First World War who had served at the front. Richard Courant belonged to this group but that did not protect him from the hatred of the Nazis against Jews in Germany. On April 26th, a call to dismiss Courant, Neugebauer, Landau, Bernays and Noether appeared in a local newspaper. Shortly after that, Courant escaped to Cambridge (UK) and later moved to New York.

Neugebauer too was considered an intolerable person—not only because he had been a member of the *Sozialdemokratische Partei Deutschlands (SPD)* (Social Democratic Party of Germany) for a short period in his young days, but, in particular, because he had spent some time in Leningrad studying Babylonian scriptures. Experiencing this animus against him and his colleagues and knowing that Bernstein, Courant and Noether were forced to leave the University of Göttingen, Neugebauer decided to resign from his post at the university. In the meantime, Harald Bohr had arranged a professorship for him in Copenhagen, which Neugebauer took up in January 1934. With the support of his wife, Grete Bruck, and Ferdinand Springer, he took the editorial office of Zentralblatt with him to Copenhagen and continued his work as editor-in-chief from there [Swe98].

In 1938, a breach occurred between Zentralblatt and its founding fathers. Wilhelm Blaschke, a member of the editorial board who was chaired in

Hamburg, had complained about the increasing number of anglophone reviews. Further, the editor Tullio Levi-Civita had been removed from the editorial board without Neugebauer's knowledge. Filled with indignation, Neugebauer resigned from his post as editor-in-chief. His colleagues Bohr, Hardy, Courant, Tamarkin and Veblen followed [Swe98]. He left Europe and emigrated to the USA. For a short time, publishing only one volume, Egon Ullrich from Gießen headed Zentralblatt.

Neugebauer was appointed to a professorship at Brown University in Rhode Island through the intervention of Roland G. D. Richardson, who, at that time was the Secretary of the American Mathematical Society. Shortly after, in 1940, Neugebauer and Richardson founded *Mathematical Reviews*, an American-based mathematical reviewing journal modelled on Zentralblatt.

The *Preußische Akademie der Wissenschaften* (Prussian Academy of Sciences) in Berlin and the *Deutsche Mathematiker-Vereinigung* (Union of German Mathematicians) took over the management of Zentralblatt in 1939. This was initiated by Ludwig Bieberbach, who was the chairman of the science division of the Prussian Academy, professor of mathematics at Berlin University, and an active member of the Nazi party NSDAP. Harald Geppert, a mathematician and also an NSDAP member, was nominated as editor-in-chief of both Jahrbuch and Zentralblatt, while Bieberbach was appointed the supervising editor. Until 1945, the editorial offices of the two journals continued working independently of each other, but

Former *Prussian Academy of Sciences*.

sometimes they shared information, scientific literature and reviews. At the end of World War II, Harald Geppert committed suicide.

The instructions of Bieberbach and Geppert, to organise the entire editorial work consistent with the Nazi ideology, were not implemented by all editors of Zentralblatt. It was not possible to prevent dismissals of Jewish staff members, but articles published by Jews or emigrants were still reviewed in Jahrbuch and Zentralblatt. For instance, almost all of the works of Wolfgang Döblin (later Vincent Doblin), the son of the Jewish-German novelist Alfred Döblin, who with his family had escaped from the Nazis to France in 1933, were reviewed in Zentralblatt. The same is true for example for articles of Courant, Einstein and Rademacher. Some reviews in English, French and Italian were also published.

6

Prof. Dr. Bieberbach   Bln.-Dahlem, den 11. Januar 1938
                                   Gelfertstr. 16

Herrn
Dr. H. Grunsky
Charlottenburg
Kaiser Friedrich Str. 59

Sehr geehrter Herr Doktor!

Besten Dank für Ihre Neujahrswünsche, die ich für Sie und Ihre Frau
erwidere. Möchte es Ihnen im neuen Jahr vor allem nun endlich be-
schieden sein, die Juden aus Ihrem Mitarbeiterstab loszuwerden. Die
Durchsicht der bisher erschienenen Hefte des Jahrgangs 1936 - von
Heften früherer Jahrgänge will ich im Augenblick nicht reden - zeigt
eine grosse Zahl an Juden vergebener Referate. Es sind zu viele,
alsdass ich eine zwingende Notwendigkeit einsehen könnte. Eine nur
bei dem betreffenden Juden vorhandene Sachkenntnis kann ich auch
nicht anerkennen. In den meisten Fällen würde z. B. ich selber ohne
weiteres in der Lage gewesen sein, das Referat zu schreiben. Soweit
ich gefunden habe, sind es die folgenden Juden, mit denen Sie Beziehungen unterhalten: Rogosinski, E. Rothe, Berwald, Freudenthal,
Mahler, Rosenthal, Behrend. Von den jüdisch versippten will ich im
Augenblick nicht reden, wiewohl auch deren Entfernung notwendig ist.
Ebenso werden Sie sich von den wegen § 175 verurteilten Referenten
trennen müssen.
Ich betone erneut, dass Sie Ihren Referentenstab nach den Richtlinien
zusammensetzen müssen, die seit dem 30. Januar 1933 für jeden Deut-
schen verbindlich sind. Sie laufen jedenfalls Gefahr, dass Ihr Handel
als mangelnder politischer Instinkt ausgelegt werde. Was soll ich im
Falle eines kürzlich von auswärts an mich herangetretenen Angriffs
auf die Akademie sagen, der man es zum Vorwurf macht, dass sie (die
Akademie) noch immer beim Jahrbuch jüdische Referenten beschäftige?
Ich kann doch darauf nur sagen, dass ich mir seit Jahr und Tag alle
Mühe gebe, zu erreichen, dass Sie die Juden abbauen, dass aber mein
Bemühen noch immer keinen vollen Erfolg hatte, weil Sie immer wieder
Gründchen finden, einen oder den anderen Juden heranzuziehen. Sie
sehen, wie Sie durch Ihr Verhalten das wohlverstandene Ansehen der
Akademie schädigen.

Heil Hitler!   (Unterschrift) Bieberbach

A letter by *Ludwig Bieberbach* from 1938, requesting *Helmut Grunsky*, Jahrbuch's main editor, to dismiss all Jews from the board of reviewers.

Erika Pannwitz, a member of the editorial board, wrote in 1947 [SS93]:

Es gibt keinen Fall, in dem eine Arbeit deshalb kurz besprochen oder ohne Referat geblieben wäre, weil der Verfasser Jude war. [3]

Also the world-famous Hermann Weyl, who had emigrated from Germany to the USA, commented in 1948 as follows [SS94]:

It is true that even during the war, the Jahrbuch continued reviewing the papers of foreign and Jewish mathematicians in an objective and decent manner.

---

3. There is no article which was only briefly reviewed, or not reviewed at all, because the author was Jewish.

A letter of Vincent Doblin, who had become a French citizen after his escape from Germany and was bound to conscript into the French army, shows the important role of Zentralblatt for him. (He published his mathematical works under the name W. Doeblin.) He wrote the following to Michel Loève in Paris from the front in the Ardennes [Bru00]:

> [...] il y a 2 choses que je voudrais bien connaître 1) mon mémoire dans le Bull. Sciences Math. sur les variables indépendantes est il paru? 2) le dernier référat du Zentralblatt de mes travaux que j'ai vu est celui de ma thèse, je vous serais reconnaissant, si le travail ne paraît pas trop long, si vous pouviez me copier les référats parus depuis.[4]

Restart in 1947

After the end of the war, Berlin was divided into four sectors, analogous to the occupation zones upon which the Allies had agreed in Yalta. Later two zones were constituted: West Berlin consisted of the sectors of the Western Allies, the Soviet sector formed East Berlin. The Soviet administration took the responsibility for the Prussian Academy of Sciences; in 1946, it was reopened under the new name *Deutsche Akademie der Wissenschaften* (German Academy of Sciences).

Some of the previous staff of Zentralblatt and Jahrbuch continued to work at the Academy but it took a while until it was formally decided how to proceed. The publication of the Jahrbuch was discontinued when its publishing house Walter de Gruyter and the Academy did not come to an agreement on a new contract for the Jahrbuch. The Academy withdrew from the former contract with effect from January 1, 1948, and the last volume, Number 68, was published in 1945. For some time, a "Jahrbuch committee" still existed at the Academy but it is not clear what its function was.

Hermann Ludwig Schmid from the Humboldt University of Berlin (formerly Friedrich-Wilhelms-Universität Berlin) became the new editor-in-chief of Zentralblatt in 1946.

Shortly after the end of the war, Zentralblatt's editorial board revived their contacts with former colleagues, inviting many of them to work again as editors or reviewers for Zentralblatt:

> Sehr verehrter Herr Kollege!
> Ihrer verehrenden Aufforderung, als Herausgeber des Zentralblattes zu zeichnen, leiste ich gerne zu, obwohl ich gar nicht weiß, was für Rechte und Pflichten das bedeutet. Alles, womit ich Ihnen helfen kann, werde ich sehr gerne machen, da ich die Wichtigkeit der Rolle des Zentralblattes in der mathematischen Literatur kenne. Aber schon allein die Tatsache, dass das Zentralblatt als erstes zu einer

---

4. [...] there are two things I'd like to know. 1) Has my article in Bull. Sciences Math. already appeared? 2) The last review of my work that I have found in Zentralblatt is about my thesis. I'd be very grateful, if it does not cause you too much effort, if you could send me copies of the reviews that have appeared since then.

left: *Hermann L. Schmid*, right: *Zentralblatt's editorial board* in 1965.

internationalen mathematischen Zusammenarbeit nach dem Kriegsende in Europa auffordert, ist so wichtig und hoffnungsvoll, dass man sich freuen kann, in dieser Arbeit teilzunehmen.
Ihr sehr ergebener Bela v. Sz. Nagy (Szeged, 30. März 1948)[5]

Sehr verehrter Herr Kollege!
In erster Linie möchte ich meiner großen Freude darüber Ausdruck geben, dass der Springerverlag das „Zentralblatt" wieder herausgeben wird. Immer wieder haben wir uns hier Gedanken über das Schicksal dieses für unsere Fachwelt so wertvollen Organs gemacht. Gerne stelle ich meine bescheidenen Mitarbeit wieder zur Verfügung—wie ich aus Erfahrung weiß, bietet die Tätigkeit als Referent durch die einlässlichere Lektüre von Spezialarbeiten aus dem eigenen Interessengebiet stets intensive Anregung, sodass ich Ihnen für die erneute freundliche Einladung dem Referentenstab beizutreten zu danken habe.
Mit vorzüglicher Hochachtung,
Ihr sehr ergebener Hugo Hadwiger (Bern, 26. Mai 1948)[6]

In particular, it should also be noted that Hanna and Bernhard Neumann, who had emigrated from Germany, started writing reviews again after the war. Hanna Neumann, née von Caemmerer, had started working for Zentralblatt in 1934. In 1938, she fled with her Jewish husband Bernhard first

5. Dear colleague! I will grant your respectful request to join the editorial board of Zentralblatt, although I do not know the rights and duties it involves. It will be my pleasure to help you as much as I can, since I know the importance of Zentralblatt for the mathematical literature. But the very fact that Zentralblatt is the first after the war in Europe to call for an international mathematical collaboration is so important and promising that one has to look forward to being part of this. Yours faithfully, Bela v. Sz. Nagy (Szeged, March 30, 1948)

6. Dear colleague! First and foremost, I want to express my happiness about the fact that Zentralblatt is again being published by the Springer publishing house. Time and again we have worried about the destiny of this journal which is so valuable for the mathematical community. With pleasure, I am offering you my cooperation as a reviewer; from experience, I know how stimulating an intense study of literature of one's own research area can be. Hence, I have to thank you for your kind invitation to rejoin the reviewer group. Yours respectfully, Hugo Hadwiger (Bern, May 26, 1948)

to England, and later to Australia. Hanna continued writing reviews until 1964, Bernhard until 2001.

The French mathematician Christian Pauc also had a special connection with Zentralblatt. On December 7, 1960, he wrote to Erika Pannwitz from Nantes:

> Die Ernennung zum Zbl.Referenten im Frühjahr 1939 verdanke ich einer Empfehlung von Herrn Haupt, und ich erinnere mich wie ich damals darauf stolz gewesen bin. Die Mitarbeit am Zbl. konnte ich im Oflag [Offizierslager] 1941–1942 fortsetzen. Dank den Bemühungen von Herrn Geppert kam ich Ende 1942 nach Berlin bei der Redaktion als beurlaubter Kriegsgefangener, Ende 1943 nach Erlangen bei Herrn Haupt und Nöbeling, wo ich bis Kriegsende als Verwalter der Assistentenstelle tätig wurde.[7]

Frédéric Roger, a second French prisoner of war, worked for the editorial team of Zentralblatt. He went to Freiburg at the end of the war, but nothing else is known about him.

The correspondence between the editors, reviewers and the authors of the reviewed articles was usually formulated in polite and respectful terms, as is, for instance, apparent from various postcards sent by authors and reviewers to Zentralblatt's editorial office.

Postcards with greetings to Zentralblatt's editorial board.

---

7. I owe my appointment as a reviewer for Zentralblatt in Spring 1939 to the recommendation of Mr. Haupt, and I remember how proud I was. I was able to continue my work for Zentralblatt in the Oflag [a prisoner of war camp for officers only] 1941–1942. Due to the efforts of Mr. Geppert I was relocated as prisoner of war first to the editorial board in Berlin at the end of 1942, and later, at the end of 1943, to Mr. Haupt and Mr. Nöbeling in Erlangen, where I worked as an assistant.

Still, some publications gave cause for conflicts and complaints. Carl L. Siegel expressed his annoyance about a review of a book on the n-body problem in very frank terms:

> Es wird behauptet, dass durch die eigenen Beiträge des Verfassers tiefere Zusammenhänge erkennbar werden und die grundlegenden Prinzipien deutlich in den Vordergrund treten. Diese Besprechung ist irreführend. In dem Buche finden sich zahlreiche unrichtige Behauptungen und mehrere wirklich schwerwiegende Fehler.[8]

In 1949, Werner Heisenberg, a physicist working within the editorial board, objected to a review because of its harsh criticism of the original work:

> Der Entwurf seines Referates, den er mir geschickt hat, ist aber so kritisch gehalten, dass er sich meiner Ansicht nach nicht für das Zentralblatt eignet, und ich möchte daher vorschlagen, dass Sie die genannte Arbeit einem anderen zuweisen.[9]

A priority dispute over the proof of an isoperimetric inequality between Alexander Dinghas and Hugo Hadwiger lasted for almost two years (1952/53); various mathematicians tried to mediate.

Zentralblatt—a cooperation between the two Germanies

In 1952, Hermann Ludwig Schmid, Zentralblatt's editor-in-chief at that time, was appointed to a professorship at the University of Würzburg. He continued his editorial work from there until he suddenly died in 1956. Erika Pannwitz became his successor. Due to her long-term working experience—she had worked as an editor for both Jahrbuch (1930–1940) and Zentralblatt (since 1947)—she clearly had great expertise in this area.

The construction of the Berlin Wall started in August 1961. The barrier cut off West Berlin from the eastern part of the city, causing many complications for Zentralblatt and its editors. While the editorial office was located in Adlershof on the premises of the German Academy of Sciences in East Berlin, Zentralblatt's publishing house Springer and, in particular, about half of its staff members, including Erika Pannwitz, were located in the western districts.

To make the communication and exchange of material between Zentralblatt's office and Springer possible, Pannwitz and two other members of staff were given special permits by the Volkspolizei, the national police of the German Democratic Republic, to enter East Berlin. This period in the history of Zentralblatt is not well documented. However, some former staff members

---

8. It is being claimed that the author's findings contribute to a deeper understanding of the problem, extracting its fundamental principles. This review is fallacious. The book under review contains false claims and several really serious mistakes.

9. The draft of the review he has sent to me is so hypercritical; I must say that it is not suitable for Zentralblatt. Hence, I suggest that you assign the named article to someone else.

have reported that at the beginning, Pannwitz used to take some publications that still needed to be reviewed with her to West Berlin. Together with other West Berlin editors, she set up a temporary editorial office within the Springer house at Heidelberger Platz. During the next four years, they continued working as an unofficial part of the editorial office, exchanging material with the eastern office on a weekly basis.

The split-up of Zentralblatt became official with the cooperation agreement between the German Academy of Sciences and the Heidelberg Academy of Sciences (Heidelberger Akademie der Wissenschaften) in 1965. The academies agreed to continue Zentralblatt with the editing duties to be shared equally, in terms of both work load and technical level, by both Berlin offices and with the printing and distribution to be done by the Springer publishing house. Walter Romberg was in charge of the eastern editorial board while Erika Pannwitz continued as editor-in-chief of the western office.

Editorial office in Jägerstr. 22/23 in Berlin, 1959: left: *Walter Romberg* with *William von Klemm* and *Herbert Benz*, right: *Erika Pannwitz* with *Elisabeth Szotowski* and *Fritz Dueball*.

It is a striking fact that this German-German cooperation continued successfully until 1977. Despite the remarkably complicated political situation in Germany, and in particular in the divided city of Berlin, Zentralblatt recovered from the drawbacks and reestablished a leading position in mathematical reviewing.

Some effects of World War II remained noticeable for a long time, like economic problems, the backlog of uncompleted reports, which was not cleared before the seventies, and the need to catch up with the Mathematical Reviews.

Erika Pannwitz retired as editor-in-chief in 1969 but she continued working as a section editor. Ulrich Güntzer from Freie Universität Berlin became her successor.

Transforming Zentralblatt into a reference database

In the seventies, the annual production of mathematical papers had reached a level which could no longer be handled by manual work. Güntzer entered several partnerships with other scientific institutions like Chemie-Information, Großrechenzentrum für die Wissenschaft (Chemistry Information, Electronic Data Processing Centre For Science) and Technische Universität (Technical University, TU) Berlin in order to benefit from their computing facilities for the editorial work at Zentralblatt. The production of the indexes of the single volumes and of the cumulated indexes had been delegated to some of the partner institutions. Bernd Wegner from TU Berlin, who became Güntzer's successor after he was offered a professorship in Tübingen in 1974, developed Güntzer's ideas of technically advancing the editorial work further. Over 35 years later, Wegner is still the editor-in-chief.

left: *Ulrich Güntzer*, right: *Bernd Wegner*.

In the seventies, the Federal Republic of Germany developed a plan to reorganise all information and documentation activities in the country. As a consequence, the Academy of Sciences of the German Democratic Republic (Akademie der Wissenschaften der DDR), formerly German Academy of Sciences and renamed in 1972, withdrew from the cooperation contract with the Heidelberg Academy and consequently refused any further collaboration. Moreover, all reviewers from the GDR had to quit their services for Zentralblatt. For about two years, Zentralblatt was solely run by the Heidelberg Academy and Springer.

In 1979 the Federal Republic of Germany established the Fachinformationszentrum Energie Physik Mathematik in Karlsruhe (FIZ Karlsruhe; Centre For Technical Information Energy Physics Mathematics) with the aim of improving and centralizing institutions dealing with scientific information and documentation in Germany. The Zentralblatt office was formally incorporated as as a subsidiary, Department Berlin, of FIZ Karlsruhe. The Heidelberg Academy remained responsible for the content of Zentralblatt, while Springer continued to publish the journal and was responsible for printing, marketing and distribution.

Between 1982 and 1985, discussions took place between the Heidelberg Academy, FIZ Karlsruhe and Springer on one side and the American Mathematical Society (AMS) on the other about a potential merger of Zentralblatt and Mathematical Reviews. Even though the negotiations failed, both review journals continue to cooperate. For example, their cooperation has led to the improvement of the Mathematics Subject Classification scheme MSC. During the 60s, the AMS developed a mathematical classification scheme called "Mathematical Offprint Service" (MOS) which was soon adopted by Mathematical reviews and somewhat later by Zentralblatt as well. The MOS scheme was revised by editors of both journals on an irregular basis. By the initiative of Bernd Wegner, in 1980, both parties concluded an agreement concerning the joint maintenance of the classification scheme, under the new name "Mathematics Subject Classification" (MSC). The agreement regulates the rules for future revisions. Since then, the MSC has been updated every ten years, the last time in 2010 [GS10].

With the support of FIZ Karlsruhe, the first release of Zentralblatt as a searchable database was established in 1989 and made accessible to the public through the database provider STN International.

With the fall of the wall in 1989, the political circumstances changed again. The Academy of Sciences of the GDR and the Institute for Applied Mathematics (Weierstraß-Institut in East Berlin) were reorganised and some former members of the East Berlin Zentralblatt editorial office resumed work for Zentralblatt. The Heidelberg Academy, FIZ Karlsruhe and Springer continued as publishers.

The revolutionary development in Information Technology and Computer Science, in particular the invention of the CD-ROM and the World Wide Web enormously helped to improve Zentralblatt's services. The first release of the database as an offline version on CD-ROM called CompactMATH was published in 1990. TEX, the ingenious typesetting system written by Donald Knuth to typeset complex mathematical formulae, was introduced in 1992. The transition of the Zentralblatt database to a service directly accessible through the internet was accomplished in 1996; it was named MATH and later ZBMATH.

Recent developments

One of the main agendas for the future has been to establish Zentralblatt as a large European infrastructure for mathematics and to enhance the Europeanisation of Zentralblatt.

Zentralblatt cooperates closely with other European mathematical institutes and scientific institutions. French support of the Zentralblatt enterprise through Cellule MathDoc (Grenoble) has been available since 1997, improving the distribution in France on the one hand and, even more importantly, developing the search software allowing a more convenient access to Zentralblatt via the internet on the other. The European Mathematical Society was invited and agreed to become involved in Zentralblatt as an additional editorial institution.

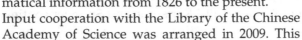

LIMES

Some streamlining in the editorial workflow was achieved by means of the close cooperation of Zentralblatt with a number of European mathematical institutions through the LIMES project (Large Infrastructure in Mathematics & Enhanced Services). The project was funded by the European Union to develop the technical prerequisites so that cooperation for a distributed input could be improved.

In 2004 Zentralblatt MATH incorporated the Jahrbuch data (1868–1942) as an extension. Moreover bibliographical data of the "Journal für die reine und angewandte Mathematik", also known as Crelle's journal, which began in 1826 were added. This makes ZBMATH the unique source of mathematical information from 1826 to the present.

Input cooperation with the Library of the Chinese Academy of Science was arranged in 2009. This group is in charge of the Chinese Mathematical Abstracts service.

Until very recently Zentralblatt was distributed in both electronic format and in print, resulting in 25 volumes of 600 pages each per year. The print service, obviously somewhat out-of-date, was discontinued in 2010. Instead Zentralblatt started offering its new print service "Excerpts from Zentralblatt MATH". In contrast to the previous print edition, not all reviews from the database are reproduced. Only a small selection suggested by the section editors are reproduced in the current print edition. Another new feature is the "Looking Back" section that allows a fresh look at classical mathematical works through contemporary reviews. For example, a review of W. Doeblin's famous *pli cacheté* has appeared in this column [BY02, Imk11].

Zentralblatt now provides a new Google-like interface accompanied by updated and improved search facilities. One important feature of ZBMATH is MathML, which allows the display of mathematical symbols and formulas in the search results directly on the screen. A comment-button has been added to the interface in order to give readers the possibility to make additions or critical (or affirmative) remarks. This service is monitored by the editors; "Dynamic Reviewing" is now possible.

Recently an author database was launched with the objective of uniquely identifying authors according to their publications together with a listing of name variations.

 In 2010 FIZ Karlsruhe/Zentralblatt MATH became partners of the "European Digital Mathematics Library" (EuDML), a project partially funded by the European Commission.

To finish this short survey about the history of Zentralblatt and to close the circle to its inception, we mention that, to honour the founder of Zentralblatt, the European Mathematical Society recently established the Otto Neugebauer Prize in the History of Mathematics. The prize will be awarded every four years at the European Congress of Mathematics for a specific  piece of work, starting in 2012. It carries a monetary award of 5000 Euros and is endowed by Springer Verlag.

Translation: Helena Mihaljevic-Brandt, Barbara O'Brien and Barry Turett

References

ABB95. Hans Aldebert, Heinz Bauer, and Jean Boclé. Christian Pauc – ein französischer Mathematiker als Wegbereiter der deutsch-französischen Freundschaft. *Mitt. Dtsch. Math.-Ver.*, 2:24–26, 1995.

Bru00. Bernard Bru. Un hiver en campagne. *C. R. Acad. Sci. Sér. I. Math.*, 331:1037–1058, 2000.

BY02. Bernard Bru and Marc Yor. Comments on the life and mathematical legacy of Wolfgang Doeblin. *Finance Stochast.*, 6:3–47, 2002.

Göt94. Heinz Götze. *Der Springer-Verlag: Stationen seiner Geschichte. Teil II, 1845–1942*. Berlin: Springer-Verlag, 1994.

GS10. Silke Göbel and Wolfram Sperber. Bibliographische Klassifikationen in der Mathematik – Werkzeug zur inhaltlichen Erschließung und für das Retrieval. *Forum der Berliner Mathematischen Gesellschaft*, 12:77–99, 2010.

GW04. Silke Göbel and Bernd Wegner. Das Zentralblatt als Zugang zur mathematischen Literatur von 1868 bis heute. *Mitt. Dtsch. Math.-Ver.*, 34(1):48–49, 2004.

Imk11. Peter Imkeller. Review of: "Wolfgang Doeblin, 'Sur l'équation de Kolmogoroff', C. R. Acad. Sci., Paris, Sér. I, Math. 331, Spec. Iss., 1059–1102 (2000)". *Excerpts from Zentralblatt MATH*, 2011/2:1–2, 2011.

16

Köt76.   Gottfried Köthe. Erika Pannwitz †. *Zentralblatt für Mathematik*, 309:3–4, 1976.

Lax03.   Peter D. Lax. Richard Courant. In *Biographical Memoirs*. *National Academy of Sciences*, volume 82, pages 78–97. The National Academies Press, 2003.

MO68.   Felix Müller and Carl Ohrtmann, editors. *Jahrbuch über die Fortschritte der Mathematik*, volume 1. Georg Reimer, 1868.

Neu31.   Otto Neugebauer. Geleitwort. In Otto Neugebauer et al., editor, *Zentralblatt für Mathematik und ihre Grenzgebiete*, page 1. Springer-Verlag, 1931.

ON63.   Otto Neugebauer. Reminiscenses on the Göttingen Mathematical Institute on the Occasion of R. Courant's 75th Birthday. The Shelby White and Leon Levy Archives Center, Institute for Advanced Study, Princeton, N.J., Otto Neugebauer Papers, Box 14, Publications vol. 11, 1963.

OR11.   John O'Connor and Edmund F. Robertson. Otto Neugebauer. MacTutor History of Mathematics, University of St. Andrews, School of Mathematics and Statistics), 2011. http://www-history.mcs.st-and.ac.uk/Mathematicians/Neugebauer.html, download: 30.05.2011.

Per52.   Oskar Perron. Harald Bohr. *Jahresber. Dtsch. Math.-Ver.*, 55:77–88, 1952.

Rei79.   Constance Reid. *Courant 1888–1972. Der Mathematiker als Zeitgenosse*. Berlin: Springer-Verlag, 1979.

Rem00.   Volker Remmert. Mathematical Publishing in the Third Reich. Springer-Verlag and the Deutsche Mathematiker-Vereinigung. *Math. Intell.*, 22(3):22–30, 2000.

RS06.   Volker Remmert and Ute Schneider. Ich bin wirklich glücklich zu preisen, einen solchen Verleger-Freund zu gewinnen. Aspekte mathematischen Publizierens im Kaiserreich und in der Weimarer Republik. *Mitt. Dtsch. Math.-Ver.*, 14(4):196–205, 2006.

RS08.   Volker Remmert and Ute Schneider. Wissenschaftliches Publizieren in der ökonomischen Krise der Weimarer Republik. Das Fallbeispiel, Mathematik in den Verlagen, B.G. Teubner, Julius Springer und Walter de Gruyter. In *Archiv für Geschichte des Buchwesens*, volume 62. München: Sauer K. G., 2008.

Sar92.   Heinz Sarkowski. *Der Springer Verlag. Stationen seiner Geschichte. Teil I*. Berlin: Springer-Verlag, 1992.

Sch83.   Norbert Schappacher. Das Mathematische Institut der Universität Göttingen 1929–1950. In München: K. G. Sauer, editor, *Universität Göttingen unter dem Nationalsozialismus*, pages 523–551. Becker, Dahms, Wegeler, 1983.

SS93.   Reinhard Siegmund-Schultze. *Mathematische Berichterstattung in Hitlerdeutschland – der Niedergang des Jahrbuches über die Fortschritte der Mathematik*. Göttingen: Vandenhoeck & Ruprecht, 1993.

SS94.   Reinhard Siegmund-Schultze. "Scientific Control" in Mathematical Reviewing and German-U.S-American Relations between the Two World Wars. *Historia Mathematica*, 21:306–329, 1994.

Swe98.   Noel M. Swerdlow. Otto E. Neugebauer. In *Biographical Memoirs. National Academy of Sciences*, volume 75, pages 214–239. The National Academies Press, 1998.

Vog99.   Annette Vogt. Von der Hilfskraft zur Leiterin. *Berlinische Monatsschrift*, 5:18–24, 1999.

Weg98.   Bernd Wegner. Berlin als Zentrum des mathematischen Referatewesens. In Heinrich G. W. Begehr, editor, *Mathematik in Berlin, Band 1*, pages 607–616. ICM 1998, Aachen: Shaker, 1998.

Weg08.   Bernd Wegner. Mathematik – Information im Wechsel der Zeiten und politischen Systeme. In *Mathematik im Blickpunkt, Wissensmanagement in der Mathematik – 140 Jahre Information zur Literatur weltweit*, pages 14–17. FIZ Karlsruhe, 2008.

Selected reviews from

Zentralblatt MATH

O. Teschke et al. (eds.), *80 Years of Zentralblatt MATH*,
DOI 10.1007/978-3-642-21172-0_2, © Springer-Verlag Berlin Heidelberg 2011

Zbl 1.00501
Courant, Richard; Hilbert, David
Methoden der mathematischen Physik. Band 1.
2. verbesserte Auflage.
Die Grundlehren der mathematischen Wissenschaften in Einzeldarstellungen. Mit besonderer Berücksichtigung der Anwendungsgebiete 12.
Berlin: Julius Springer. XIV, 469 S. und 26 Abb. (1931).

Dem Physiker war es fast immer bewußt, daß die Stärke seiner Wissenschaft und ihre rasche Entwicklung mit auf der Anwendung der mathematischen Sprache und der mathematischen Methoden beruht. Für die Physik besonders wichtig und zum großen Teil aus physikalischen Fragestellungen entstanden sind die Theorien der gewöhnlichen und partiellen Differentialgleichungen, die Variationsrechnung, die Entwicklungen von Funktionen und der Gebrauch bestimmter, den geometrischen Eigenschaften vieler

Author profile:

Courant, Richard

Spellings: Courant, Richard [82] Courant, R. [76]
Author-Id: courant.richard
Publications: 158 including 59 Book(s) and 87 Journal Article(s)

MSC 2010

17	**00** · General mathematics
10	**26** · Real functions
10	**34** · Ordinary differential equations (ODE)
10	**35** · Partial differential equations (PDE)
7	**01** · History; biography

more ...

Journals

17	Proceedings of the National Academy of Sciences of the United States of America
6	Mathematische Annalen
5	Communications on Pure and Applied Mathematics
4	Jahresbericht der Deutschen Mathematiker-Vereinigung (DMV)
3	Acta Mathematica

more ...

Co-Authors

15	Hilbert, David M.
14	Robbins, Herbert E.
7	John, Fritz
4	Friedrichs, Kurt Otto
3	Lax, Peter D.

more ...

Publication Years

physikalischer Systeme besonders angepaßter Funktionen. Diese Fragen behandelt der vorliegende Band unter Gesichtspunkten, die Einheitlichkeit, Wichtigkeit und Schönheit der Theorien klar hervortreten lassen. Den Mittelpunkt der Darstellung bildet die Theorie der linearen Integralgleichungen.

Um die Analogie der linearen Integralgleichungen zu den linearen Gleichungen der Algebra und die Analogie vollständiger orthogonaler Funktionssysteme zu den Komponenten eines Vektors aufzeigen zu können, wird zunächst die Algebra linearer Transformationen behandelt. Es folgt das Problem der Reihenentwicklung willkürlicher Funktionen als Verallgemeinerung der Entwicklung in Fouriersche Reihen. Die Grundtatsachen der Variationsrechnung führen über das Hamiltonsche Prinzip der Mechanik zu den Schwingungs- und Eigenwertproblemen der Physik. Bestimmte durch Eigenwertprobleme definierte Funktionen, die für den Physiker besonders wichtig sind (Besselsche Funktionen und Kugelfunktionen) bilden den Abschluß. Im Vorwort zur 1. Auflage klagt R. Courant über eine Lockerung der Beziehungen zwischen Mathematiker und Physiker. Außer den von ihm angegebenen, in der Entwicklung der Mathematik liegenden Gründen war im letzten Jahrzehnt auch noch der Umstand schuld, daß ein wichtiger Zweig der theoretischen Physik (die Quantentheorie) sich auf seine Grundlagen besinnen mußte und vor Sicherung dieser Grundlagen nicht an einen Ausbau mit mathematischen Methoden denken konnte. Dieser Abschnitt der

Author profile:

Hilbert, David M.

Spellings: Hilbert, D. [163] Hilbert, David [52] Hilbert,D. [1] Hilbert, David M. [1]
Author-Id: hilbert.david-m
Publications: 216 including 74 Book(s) and 137 Journal Article(s)

MSC 2010

21	01 · History; biography
18	11 · Number theory
11	51 · Geometry
9	00 · General mathematics
8	03 · Mathematical logic

more ...

Journals

17	Mathematische Annalen
8	Jahresbericht der Deutschen Mathematiker-Vereinigung (DMV)
6	Acta Mathematica
4	L'Enseignement Mathématique
1	Abhandlungen aus dem Mathematischen Seminar der Universität Hamburg

more ...

Co-Authors

15	Courant, Richard
10	Bernays, Paul
8	Ackermann, Wilhelm
4	Cohn-Vossen, Stefan
1	Cohn-Vossen, St.

more ...

Publication Years

Physik wurde nun gerade in den Jahren nach Erscheinen der 1. Auflage des vorliegenden Bandes mit erstaunlicher Schnelligkeit überwunden. Und seitdem werden gerade in der Quantentheorie (sei es bei den Methoden, die Schrödinger einführte, sei es bei der Benutzung von Matrizen und Operatoren) die Methoden, die dieses Buch behandelt, ausgiebig benutzt. Die rasche Entwicklung dieses neuesten Zweiges der theoretischen Physik wäre nicht möglich gewesen, wenn die mathematischen Methoden nicht bereitgelegen hätten und Mathematiker auf ihre Wichtigkeit hingewiesen hätten. Es gibt sicher manchen theoretischen Physiker, dem die aktive oder rezeptive Teilnahme an der genannten Entwicklung möglich wurde durch Hilbertsche oder Courantsche Vorlesungen oder durch das vorliegende Buch. Die Abweichungen der 2. Auflage von der 1. sind zum Teil wohl mit Rücksicht auf die Entwicklung der Physik erfolgt. So sind die Hermiteschen Matrizen und die unitären Transformationen mitbehandelt, komplexwertige Funktionen sind mitberücksichtigt, auf die Sturm-Liouvilleschen Differentialgleichungen ist ausführlicher eingegangen, Probleme mit kontinuierlichem Spektrum von Eigenwerten, die Schrödingersche Differentialgleichung und die Behandlung "gestörter" Differentialgleichungen sind zugefügt. Auch sonst sind im einzelnen zahlreiche Abänderungen und Ergänzungen erfolgt. Besonders der Abschnitt über Variationsrechnung enthält einiges Neue. Durch Kürzungen an anderen Stellen ist erreicht, daß der Umfang des ganzen Buches nicht wesentlich gewachsen ist.

F. Hund (Leipzig)

Zbl 5.36204
Grunsky, Helmut
Neue Abschätzungen zur konformen Abbildung ein- und mehrfach zusammenhängender Bereiche.
Schr. Math. Semin. u. Inst. Angew. Math. Univ. Berlin 1, 95-140 (1932).

Die Grundlage der Untersuchung bildet eine allgemeine Ungleichung, die für sämtliche Funktionen $f(\zeta) = re^{i\phi}$ gilt, welche in einem endlich vielfach zusammenhängenden Bereich $\mathcal{B}$ meromorph sind und dort endlich viele Null- und Unendlichkeitsstellen besitzen. Sie gibt eine untere Schranke für das Integral $\int_{\mathcal{B}} \log r\,d\phi$, erstreckt über den Rand $\mathcal{R}$ von $\mathcal{B}$ in positivem Sinne. Diese enthält von $f(\zeta)$ die ersten Laurentkoeffizienten in den Nullstellen und Polen der Funktion und die Umlaufszahlen (Gesamtzuwachs des Arkus) der Funktion längs der einzelnen Randkomponenten. Neben diesen Größen enthält die Schranke endlich viele Parameter, die von gewissen von $\mathcal{B}$ und der Lage der Nullstellen und Pole abhängigen Hilfsfunktionen herrühren. Durch dem jeweils vorliegenden Problem angepaßte Spezialisierung derselben wird eine Reihe von neuen Abschätzungen erhalten, von denen einige hervorgehoben seien.

1. $\mathcal{B}$ enthalte $\zeta = 0$ und $\zeta = \infty$. $f(\zeta)$ vermittle eine schlichte Abbildung von $\mathcal{B}$, und es sei $f(\infty) = \infty$, $f'(\infty) = 1$. Dann gibt es zu jedem ζ aus $\mathcal{B}$ zwei Zahlen $r(\zeta)$, $m(\zeta)$, von denen die erste positiv ist, so daß

$$\log f'(\zeta) - m(\zeta) \le r(\zeta)$$

gilt. Unter $\log f'(\zeta)$ ist der bei $\zeta = \infty$ verschwindende Zweig zu verstehen. Zu jedem Wert $m(\zeta) + \tau r(\zeta)$ $|\tau| = 1$ gibt es eine Funktion, so daß $\log f'(\zeta) = m(\zeta) + \tau r(\zeta)$ ist. Sie ist bis auf eine additive Konstante bestimmt und vermittelt die Abbildung auf die längs gewisser Bögen von logarithmischen Spiralen mit dem asymptotischen Punkt $f(\zeta)$ aufgeschlitzte Ebene.

Einen noch allgemeineren Satz hat *H. Grötzsch* bewiesen [Ber. Verh. Sächs. Akad. Leipzig 84, 15–36 (1932; Zbl 5.06802); Ber. Verh. Sächs. Akad. Leipzig 84, 3–14 (1932; Zbl 5.06801)].

2. Bildet $s(\zeta)$ den Einheitskreis $\zeta < 1$ auf einen schlichten Bereich ab, der $s = \infty$ nicht enthält, so ist, falls $s(0) = 0$, $s'(0) = 1$ ist,

$$\left| \log \frac{s(\zeta)}{\zeta} + \log(1 - |\zeta|^2) \right| \le \log \frac{1 + |\zeta|}{1 - \zeta},$$

$$\left| \log \frac{s'(\zeta)\zeta}{s(\zeta)} \right| \le \log \frac{1 + |\zeta|}{1 - \zeta}, \quad |\log s'(\zeta) + \log(1 - |\zeta|^2)| \le 2 \log \frac{1 + |\zeta|}{1 - \zeta}.$$

Die Logarithmen bedeuten jedesmal den für $\zeta = 0$ verschwindenden Zweig. Alle Schranken sind scharf.

Author profile:

Grunsky, Helmut

Spellings: Grunsky, Helmut [44] Grunsky, H. [14]
Author-Id: grunsky.helmut
Publications: 58 including 4 Book(s) and 48 Journal Article(s)

MSC 2010

13	**30** · Functions of a complex variable
2	**01** · History; biography
1	**26** · Real functions
1	**47** · Operator theory
1	**53** · Differential geometry

more ...

Journals

13	Jahresbericht der Deutschen Mathematiker-Vereinigung (DMV)
11	Mathematische Zeitschrift
4	Deutsche Mathematik
4	Mathematische Annalen
2	MNU. Der Mathematische und Naturwissenschaftliche Unterricht

more ...

Co-Authors

2	Ahlfors, Lars Valerian
1	Jenkins, James A.

Publication Years

In einem Schlußparagraphen wird das Extremumproblem gelöst, unter allen in 58 regulären Funktionen, die an zwei Stellen vorgeschriebene Werte annehmen, diejenige zu finden, für die der Flächeninhalt des Bildes ein Minimum wird. Für einfach zusammenhängende Bereiche ist es bereits früher von Kubota behandelt worden.　　　　　　　　　　*K. Loewner (Prag)*

Zbl 7.07501
Severi, Francesco
Über die Grundlagen der algebraischen Geometrie.
Abh. Math. Semin. Hamb. Univ. 9, 335-364 (1933).

Es handelt sich in dieser reichhaltigen Arbeit vor allem um eine rein algebraisch-geometrische Definition der Schnitt punktsmultiplizität, um den Beweis des verallgemeinerten Bézoutschen Theorems und des Prinzips der Erhaltung der Anzahl. Zunächst wird die Ordnung einer Mannigfaltigkeit M_k im projektiven Raum S_r, definiert durch Zurückführung auf den Begriff der Ordnung einer Hyperfläche, indem die M_k aus einem allgemeinen Punkt auf eine S_{k+1} projiziert wird. Die Ordnung von M_k ist auch gleich der Anzahl ihrer Schnittpunkte mit einer allgemeinen S_{n-r}, des S_r. Sind nun zwei Mannigfaltigkeiten V_k, W_{n-k} von den Ordnungen m, n in S_r gegeben und bildet man aus W die Familie aller Mannigfaltigkeiten $\overline{W}$, die durch projektive Transformationen in W übergehen können (oder eine noch umfassendere irreduzible Familie), so wird das verallgemeinerte Bézoutsche Theorem folgendermaßen formuliert: Die allgemeine Mannigfaltigkeit W schneidet Y in genau $m \cdot n$ Punkten. Die Vielfachheit eines isolierten Schnittpunktes P von Y, W ist gleich der Anzahl der Schnittpunkte von V und $\overline{W}$, die gegen P konvergieren, wenn $\overline{W}$ in der Familie gegen W strebt. Haben V, W nur endlich viele Punkte gemein, so ist die Summe der Vielfachheiten dieser Schnittpunkte gleich $m \cdot n$. Der Beweis benutzt das Chaslessche Korrespondenzprinzip (mit der Zeuthenschen Regel), angewandt auf die Punktepaare A, A', bei denen A auf Y, A' auf W (bzw. $\overline{W}$) liegt und die Verbindungslinie AA' durch einen festen Punkt S geht. Dabei zeigt sich noch: Der Schnittpunkt P zählt einfach, wenn Y und W durch P einfach hindurchgehen und keine gemeinsame Tangente haben.

Die obige Multiplizitätsdefinition stimmt mit der vom Referenten [*B.L. van der Waerden*, "Eine Verallgemeinerung des Bézoutschen Theorems," Math. Ann. 99, 497–541 (1928; JFM 54.0141.01)] gegebenen überein. Sie versagt, wenn V_k, W_{r_k} nicht dem S_r, sondern einer Mannigfaltigkeit M_r des Raumes S_d angehören, wobei P kein singulärer Punkt von M_r sein darf. Verfasser gibt für diesen Fall statt der topologischen Definition von Lefschetz und dem Referenten die folgende geometrische Definition: Man betrachte eine Mannigfaltigkeit $H = V + V'$ in M, derart, daß Y' nicht durch P geht und H in einer kontinuierlichen Familie enthalten ist, deren allgemeine Mannigfaltigkeit $\overline{H}$ mit W in der Umgebung von P nur einfache Schnitte besitzt, und definiere die Vielfachheit von P wie oben durch den Grenzübergang $\overline{H} \to H$

24

Author profile:

Severi, Francesco

Spellings: Severi, F. [215] Severi, Francesco [132] Severi, Fr. [6]
Author-Id: severi.francesco
Publications: 353 including 33 Book(s) and 303 Journal Article(s)

MSC 2010

9	14 · Algebraic geometry
1	00 · General mathematics

Journals

22	Atti della Accademia Nazionale dei Lincei
20	Annali di Matematica Pura ed Applicata. Serie Quarta
20	Comptes Rendus de l'Académie des Sciences. Paris
18	Rendiconti del Seminario Matematico delle Facoltá di Scienze della R. Universitá di Roma, II. Serie
12	Rendiconti del Circolo Matematico di Palermo

more ...

Co-Authors

5	Enriques, Federico
3	Segre, Beniamino
2	Dragoni, Giuseppe Scorza
1	Aliotta, Antonio
1	Armellini, Giuseppe

more ...

Publication Years

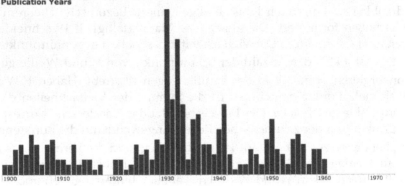

kann. H kann z. B. durch Projektion von Y aus einem allgemeinen S_{d-r-1} des Raumes S_r erhalten werden. Die Vielfachheit ist unabhängig von der Wahl von H, symmetrisch bei Vertauschung von Y und W und invariant bei algebraischen Transformationen, die in der Umgebung von P eineindeutig sind. Sie ist 1, wenn Y und W in P keine Tangente gemeinsam haben. Der Begriff wird noch auf den Fall zweier Mannigfaltigkeiten V_h, W_k mit $h+k < r$ ausgedehnt. Schließlich wird die vom Verf. 1912 gegebene Formulierung des Prinzips der Erhaltung der Anzahl näher begründet und der Begriff der virtuellen Schnittpunktszahl zweier Mannigfaltigkeiten V_k, W_{r-k} in M_r erklärt. *Bartel van der Waerden (Leipzig)*

Zbl 10.09701
Neugebauer, Otto
Vorlesungen über Geschichte der antiken mathematischen Wissenschaften. Band 1. Vorgriechische Mathematik.
Die Grundlehren der mathematischen Wissenschaften in Einzeldarstellungen mit besonderer Berücksichtigung der Anwendungsgebiete. 43.
Berlin: Julius Springer. XII, 212 S., 61 Fig. (1934).

Die Absicht des auf drei Teile berechneten Werkes besteht einerseits in der Darstellung der vorgriechischen Mathematik, andererseits in der Rekonstruktion der voreuklidischen Periode der griechischen sowie der klassischen griechischen Mathematik und der babylonischen und griechischen Astronomie. Der vorliegende erste Band betrifft die altorientalische, d. h. babylonische und ägyptische Mathematik. Das erste Kapitel behandelt die

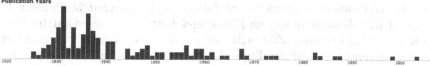
Tabellentexte, die die Grundlage der babyionischen Rechentechnik bilden; mittels einer geometrischen Darstellung der regulären Zahlen und ihrer Reziproken, welche dieselbe Unbestimmtheit besitzt wie ihre positionelle Schreibweise, wird nachgewiesen, daß die Reziprokentabellen durch fortgesetzte Zwei-, Drei- und Fünfteilung ans der am meisten vorkommenden

Normaltabelle abgeleitet worden sind. Bei der Behandlung der Multiplikationstabellen ergibt sich das Resultat, daß dieselben ursprünglich angelegt sind als Hilfsmittel zur sexagesimalen Darstellung von Vielfachen von Stammbrüchen; nach Hinzufügung einer Tabelle mit der Kopfzahl 7 (der einzigen nichtregulären Zahl kleiner als 10) konnten sie dann auf Grund des positionellen Charakters des Ziffernsystems als Multiplikationstabellen benutzt weiden. Das zweite Kapitel erläutert nach einer chronologischen und geographischen Übersicht der in Rede stehenden Kulturen das Prinzip der Keilschrift, deren Ausdrucksmöglichkeiten für die mathematische Entwicklung von wesentlicher Bedeutung gewesen sind. Die Benutzung des sumerischen Schriftsystems in der von der sumerischen strukturell grundverschiedenen akkadischen Sprache führt zu Umwandlungsprozessen, die u. a. auch die Möglichkeit einer Symbolschrift zur Darstellung von mathematischen Operationen und dadurch einer Algebra ergeben. Das Kapitel schließt mit einer kurzen Besprechung der ägyptischen Schrift. Das dritte Kapitel ist der Behandlung der babylonischen und ägyptischen Zahlensysteme gewidmet; ausführlich werden die Individualbezeichnungen der kleineren Bruchzahlen in Sprache und Schrift bei verschiedenen Völkern behandelt. Es folgt eine Diskussion der Entwicklungsgeschichte des Sexagesimalsystems, in der die bekannte Hypothese des Verf. geschildert wird: das Bedürfnis nach Aneinanderfügung von zwei ursprünglich isoliert stehenden Maßgruppen für Gewichte, wobei die natürlichen Bruchteile der größeren Einheit als ganzzahlige Multiple der kleineren erscheinen sollten, führt zu dem Verhältnis 1: 60 für Schekel : Mine; die konkreten Maßbruchteile verwandeln sich in Bruchbezeichnungen schlechthin; die sexagesimale Struktur des Gewichtssystems wird auf andere Gebiete übertragen, und es ergibt sich eine allgemeine sexagesimale Bezeichnungsweise für Bruchteile und Vielfache. Der Verzicht auf ausdrückliche Nennung der Einheit liefert unmittelbar die Stellenwertsbezeichnung, wodurch der positionelle Charakter des Systems erklärt wird. Im vierten Kapitel wird ausgeführt, wie auch ihrerseits die ägyptische Mathematik aufs tiefste beeinflußt worden ist von der Struktur ihres Zahlensystems und der damit verknüpften Rechentechnik; das führt zu einer gründlichen Auseinandersetzung mit dem Problem der 2/n-Tabelle; außerdem bringt das Kapitel eine allgemeine Charakteristik, der ägyptischen Mathematik an der Hand von arithmetischen und geometrischen Problemen; speziell wird dabei eingegangen auf das Problem M 10, das Struve als Berechnung der Halbkugeloberfiäche gelesen hat. Im fünften Kapitel wird in großen Zügen die babylonische Mathematik geschildert, die, wie gezeigt wird, in der Entwicklung ihrer Algebra eine vor Beginn der neueren Mathematik nicht wieder errungene Stufe erreicht hat. Ein abschließender Paragraph, in dem die allgemeine Problemlage skizziert wird, führt zu den beiden noch zu erwartenden Teilen des Werkes hinüber.

E. J. Dijksterhuis (Oisterwijk, Holland)

Zbl 13.07904
Alexandroff, Paul; Hopf, Heinz
Topologie. Band 1. Grundbegriffe der mengentheoretischen Topologie.
Topologie der Komplexe. Topologische Invarianzsätze und anschliessende
Begriffsbildungen. Verschlingungen im n-dimensionalen euklidischen
Raum. Stetige Abbildungen von Polyedern.
(Die Grundlehren der mathematischen Wissenschaften in Einzeldarstel-
lungen mit besonderer Berücksichtigung der Anwendungsgebiete. 45.)
Berlin: Julius Springer. XIII, 636 S., 39 Abb. (1935).

"Die Verff. haben sich die Aufgabe gestellt, in lückenloser Darstellung, ohne
die Allgemeinheit und Abstraktion der Begriffsbildung zu scheuen, die
grundlegenden Resultate einer erfolgreichen Periode in der Entwicklung
der Topologie – einer Periode, die mit Poincaré beginnt und in den Arbeiten
von Brouwer, Alexander u. a. zur vollen Geltung gekommen ist, zusammenzufassen." Das Ziel des Werkes ist nicht eine "Darstellung der ganzen

Author profile:

Alexandroff, Pavel Sergejewitsch

Spellings: Alexandroff, P. [96] Alexandroff, P.S. [32] Alexandroff, Paul [20] Alexandroff, Pavel Sergejewitsch [1] Alexandroff, P. S. [1]
Author-Id: alexandroff.pavel-sergejewitsch
Publications: 150 including 20 Book(s) and 128 Journal Article(s)

MSC 2010

4	**54** · General topology
3	**00** · General mathematics
3	**26** · Real functions
3	**55** · Algebraic topology
2	**01** · History; biography

more ...

Journals

23	Comptes Rendus de l'Académie des Sciences. Paris
13	Fundamenta Mathematicae
13	Mathematische Annalen
9	Uspekhi Matematicheskikh Nauk
8	Comptes Rendus (Doklady) de l'Académie des Sciences de l'URSS, Nouvelle Série

more ...

Co-Authors

6	Urysohn, P. †
5	Hopf, Heinz
4	Chintschin, A.J.
4	Markuschewitsch, A.I.
4	Pontrjagin, Leon

more ...

Publication Years

1910   1920   1930   1940   1950   1960   1970   1980   1990

Topologie", sondern die "Vorstellung der Topologie als eines Ganzen" zu
geben (aus dem Vorwort). Dieser synthetische Standpunkt, im Gegensatz zu
dem früher üblichen Unterscheiden zwischen der kombinatorischen und der
mengentheoretischen Topologie, ist einer der charakteristischen Züge des
Buches, der von Topologen beider Richtungen zweifellos mit Anerkennung
begrüßt wird: es handelt sich ja um Aufstellung einer einheitlichen Theorie,

28

welche die bisherigen Ergebnisse zweier so entfernt liegender Gebiete der Mathematik, wie die Punktmengenlehre und die algebraische (gruppentheoretische) Theorie der Komplexe, in sich vereinigt. Zur Verwirklichung dieses Zieles sind beide Verff., denen die genannten Gebiete zahlreiche grundlegende Ergebnisse verdanken, ganz besonders geeignet.

Der Hauptinhalt dieses ersten von den drei Bänden, aus welchen das Werk bestehen soll, ist – laut Angabe der Verff. – die Topologie der n-dimensionalen allgemeinen Polyeder. Anwendungen auf allgemeine topologische Räume werden nur gelegentlich angegeben (systematische Anwendungen finden voraussichtlich im Bd. II, bei Betrachtung kompakter Räume, ihren Platz, während andererseits eine ausführliche Untersuchung spezieller Polyeder, namentlich der Mannigfaltigkeiten, für Bd. III vorbehalten ist).

Nach einem kurzen, über Entstehung und Ziel des Werkes berichtenden Vorwort folgt die eigentliche Einleitung, in welcher die Hauptlinien der Entwicklung und des gegenwärtigen Zustandes der Topologie sowie ihre Zusammenhänge mit anderen Zweigen der Mathematik näher besprochen werden.

Author profile:

Hopf, Heinz

Spellings: Hopf, H. [74] Hopf, Heinz [37]
Author-Id: hopf.heinz
Publications: 111 including 6 Book(s) and 95 Journal Article(s)

MSC 2010

4	55 · Algebraic topology
3	53 · Differential geometry
2	01 · History; biography
1	54 · General topology
1	57 · Manifolds and cell complexes

Journals

25	Commentarii Mathematici Helvetici
14	Mathematische Annalen
7	Jahresbericht der Deutschen Mathematiker-Vereinigung (DMV)
5	Mathematische Zeitschrift
4	Fundamenta Mathematicae

more ...

Co-Authors

5	Alexandroff, Pavel Sergejewitsch
4	Rinow, Willi
4	Samelson, Hans
3	Pannwitz, Erika
2	Pontrjagin, Leon

more ...

Publication Years

Der Band zerfällt in vier Teile. Der erste Teil hat die Grundbegriffe der mengentheoretischen Topologie zum Gegenstand. Obwohl die mengentheoretischen Betrachtungen in diesem Bande bloß eine Hilfsrolle spielen, ist der genannte Teil nicht vom Standpunkt unmittelbarer Anwendungen auf die Polyedertopologie, sondern vielmehr als eine planmäßige, wenn auch beschränkte Darstellung der Theorie topologischer (d. h. auf der Bildung der abgeschlossenen Hülle $\overline{A}$ als dem Grundbegriffe gestützter), metrischer und kompakter Räume bearbeitet worden. Der zweite Teil ist den Grundlagen der kombinatorischen Topologie (Topologie der Komplexe) gewidmet. Sein Anfangskapitel handelt von Polyedern und ihren Zellenzerlegungen und bezieht sich auf den (n-dimensionalen) Euklidischen Raum: Zellen sind gewisse Punktmengen dieses Raumes und Zellenkomplexe gewisse Mengen von Zellen (also Mengen von Punktmengen). Dagegen haben weitere Kapitel dieses Teiles bereits einen durchaus abstrakten Charakter: Eckpunktbereiche sind Mengen von ganz beliebigen Elementen, Simplex ist eine endliche Menge von Eckpunkten; absoluter Komplex heißt eine (gewissen Bedingungen genügende) Menge von Simplexen und algebraischer Komplex wird als eine in bezug auf orientierte Simplexe lineare Form, mit einer abelschen Gruppe als Koeffizientenbereich, erklärt. Für die so definierten Begriffe werden diejenigen von Rand, Zykel, Homologie (in bezug auf einen doppelten Koeffizientenbereich den einen für berandende Zykeln und den zweiten für berandete Komplexe), Zusammenhang usw. eingeführt, ferner Bettische Gruppe in bezug auf verschiedenartige Koeffizientenbereiche (mit Untersuchung ihrer wechselseitigen Beziehungen), Bettische Zahlen (nebst Additions- und Produktsätzen), kombinatorische Zellen (in Anlehnung an den Begriff der Zellenzerspaltung), Torsion (übrigens ohne Gebrauch von Matrizen für Torsionskoeffizienten) u. a.

In den beiden letzten Teilen des Bandes kehren Verff. zum Euklidischen Raum zurück, um nun den so aufgebauten abstrakt-algebraischen Apparat auf ihn anzuwenden. Das Vorgehen erfolgt also nach dem von Verff. formulierten Programm: "Wir entnehmen die Grundbegriffe...dem konkreten und am Zufälligen haftenden Material der Polyeder, lassen sie dann die läuternde Wirkung der stärksten Abstraktion erfahren, um ein Werkzeug zu erhalten, welches nachher wiederum auf die konkrete geometrische Wirklichkeit – im allgemeinen wie im speziellen – angewendet werden soll" (S. 154). So beginnt der dritte Teil mit dem Kapitel über simpliziale Abbildungen, Approximationen beliebiger stetiger Abbildungen durch dieselben, Invarianz der Dimensionszahl, Theorie stetiger Komplexe, Borsukscher Begriff des Retraktes. Das zweite Kapitel betrifft kanonische Verschiebungen, einen mit dem bekannten Überführungssatz von Alexandroff eng verbundenen Begriff. Dieser Satz über Abbildungen von Kompakta in "Nerven" ihrer Zerlegungen erlaubt die kombinatorischen Methoden auf die Theorie kompakter Räume anzuwenden, was bereits in diesem und dem danach folgenden Kapitel zum Ausdruck kommt. Die beiden Kapitel gehören vollständig zur "einheitlichen" Topologie: keine Trennung – etwa in mengentheoretische und kombinatorische Sätze – wäre hier möglich. Bettische Zahlen von Nerven eines gegebenen Kompaktums K erlauben sog. Bettische N–Zahlen von K zu definieren. Insbesondere ist die $(n-1)$-te Bettische N-Zahl einer kompakten Teilmenge K des n-dimensionalen Euklidischen Raumes $\mathbb{R}^n$ von Bedeutung: Auf Grund des Zerlegungssatzes unterscheidet sich nämlich diese Zahl um 1 von der Anzahl der Komponenten von $\mathbb{R}^n \setminus K$, was (selbst die quantitative) Invarianz des Schnittes, Jordan–Brouwerschen Satz, Gebietsinvarianz sowie eine Reihe von Sätzen über irreduzible Schnitte, Cantorsche Mannigfaltigkeiten usw. ergibt. Außerdem enthält dieser Teil des Bandes direkte Invarianzbeweise für Dimensionszahlen und Gebiete sowie Beweise für Invarianz der Bettischen Gruppen und für Äquivalenz zwischen Zerschneidung von $\mathbb{R}^n$ und Existenz von sog. wesentlichen Abbildungen auf die Sphäre S^{n-1}. Alle diese Invarianzsätze "ermöglichen die Übertragung der Begriffe und Methoden aus der Topologie der Komplexe auf Polyeder; sie sind somit das Fundament der Topologie der Polyeder" (S. 409).

Der vierte Teil beginnt mit dem Kapitel über Verschlingungstheorie (Schnittzahlen, Verschlingungszahlen, Existenz von verschlungenen Zykeln in bezug auf verschiedene Koeffizientenbereiche). Auf Grund dieser Theorie wird der "nach heutigem Zustand unserer Kenntnisse umfassendste aller Sätze über die topologische Lage von krummen Polyedern im $\mathbb{R}^n$, namentlich der Alexandersche Dualitätssatz, bewiesen. Das zweite Kapitel handelt über Brouwerschen Abbildungsgrad und Kroneckersche Charakteristik. Der erste dieser Begriffe wird an Verschlingungszahl von Zykel mit Punkt und seine Theorie auf Verschlingungstheorie gestützt. Unter den zahlreichen Anwendungen dieses Begriffes sind insbesondere die Anwendungen auf Vektorfelder, Fixpunktsätze und antipodentreue sowie wesentliche Abbildungen zu erwähnen. Im darauffolgenden Kapitel über Homotopie von Abbildungen werden Sätze angegeben, die aus den Homologieeigenschaften von Abbildungen ihre Homotopieeigenschaften herzuleiten erlauben, z. B. der bekannte Satz von Hopf über Abbildungen von Polyedern in Sphären. Es werden ferner Homotopieeigenschaften von Abbildungen auf Kreislinie (im Zusammenhang mit dem Verschwinden der ersten Bettischen Zahl) sowie die Charakterisierung der Geschlossenheit und des Randes von Polyedern durch Deformationseigenschaften eingehend betrachtet. Das letzte (XIV.) Kapitel ist den mit Homologieeigenschaften verknüpften Fixpunktsätzen gewidmet. Mit Hilfe des Begriffes von Spuren der Autohomomorphismen definieren Verff. die "Lefschetzsche Zahl einer Abbildung" und gelangen zur Hopfschen Verallgemeinerung der Euler–Poincaréschen Formel und deren zahlreichen Anwendungen. Weitere Sätze betreffen den Index von Fixpunkten, die regulären Fixpunkte, ihre algebraische Anzahl und die Richtungsfelder in geschlossenen Mannigfaltigkeiten.

Zwei Anhänge über Abelsche Gruppen und konvexe Zellen im $\mathbb{R}^n$ schließen den Band ab.

Durch präzise Formulierung der Definitionen und Einfachheit der Beweise entspricht das Buch sämtlichen Anforderungen an logische Strenge, aber zugleich an geometrische Anschaulichkeit, dank vielen lehrreichen Bemerkungen, meisterhaft gewählten Beispielen und sorgfältigen Bildern. Allgemeinheit der Fassung und Vollständigkeit der Darstellung (die sich bis auf neueste, teilweise sogar noch unveröffentlichte Ergebnisse erstreckt) verleihen dem Buche nicht nur den Wert eines ausgezeichneten Lehrbuches über den gegenwärtigen Zustand der betrachteten Teile der Topologie, sondern auch die Bedeutung eines für weitere Untersuchungen aussichtsvollen Ausgangspunktes. *Kazimierz Kuratowski (Warszawa)*

Zbl 14.16304
Nevanlinna, Rolf
Eindeutige analytische Funktionen.
(Die Grundlehren der mathematischen Wissenschaften in Einzeldarstellung mit besonderer Berücksichtigung der Anwendungsgebiete. 46.)
Berlin: Julius Springer. VIII, 353 S., 24 Abb. (1936).

Dans cet ouvrage, l'éminent analyste d'Helsingfors expose une partie des beaux resultats obtenus ces dernières années par lui, ses élèves et quelques autres, dans l'étude de la distribution circulaire des valeurs (ou des éléments superficiels) prises par les fonctions uniformes $w = w(z)$ dont le domaine naturel d'existence est simplement connexe. w décrit une surface de Riemann S, simplement connege, ouverte dès que $w(z)$ n'est pas une fraction rationnelle. L'auteur porte son attention sur les propriétés de S, plus qu'on ne Pavait fait jusqu'ici dans les ouvrages consacrés à cette théorie. Ses remarquables travaug sur les fonctions méromorphes qui l'ont conduit à une conception nouvelle de la notion de valeur exceptionnelle, celle de défaut, l'ont amené à construire des classes de fonctions admettant des valeurs dont la somme des défauts atteint le maximum 2. Ses recherches et celles de ses élèves furent ainsi orientées dans des directions nouvelles: étude des fonctions pour lesquelles les singularités de S sont en nombre fini; comparaison de la relation des défauts et de la formule de Riemann donnaut la ramification moyenne d'une surface S fermée, formation de criteres permettant de reconnaître si une surface S ouverte est du type parabolique (représentable conformément sur le plan simple pointé) ou hyperbolique (représentable sur un cercle de rayon fini). Les succès obtenus dans cette dernière voie, les difficultes quelle presente encore, ont conduit l'auteur à donner à cette recherche des critères de type la place centrale dans cet exposé. C'est vers ce but que tont converge et c'est ce qui explique le plan de l'ouvrage et le choix des matières esposées et aussi que ce livre dans lequel l'auteur ne presente que des questions qu'il a fortement marquées de son empreinte et donne pour la premiére fois en details sa théorie de la mesure harmonique soit tout à la fois didactique et profondement original. Voici le contenu des chapitres:

I. Transformation conforme d'un cercle en, lui-même; ses invariants; premiere introduction de la mesure harmonique. Enoncés généraux sur la représentation conforme; cas des surfaces de Riemann S. Définition de la surface universelle de recouvrement d'un domaine multiplement

Author profile:

connexe. Surfaces S admettant seulement p singularités logarithmiques (fonctions modulaire et automorphes).

II. Integrale de Poisson et resolution du probleme de Dirichlet pour le cercle unité Γ: Mesure harmonique d'un arc de Γ en un point interieur (c'est la valeur en ce point de la fonction harmonique bornée egale à 1 sur cet arc et à 0 sur le reste de Γ). Definition generale de la mesure harmonique d'un arc d'un contour au moyen de la fonction harmonique; lignes d'egale mesure harmonique.

III. Invariance de la mesure harmonique dass la rep. conforme. Méthode de la mesure harmonique (c'est à'dire comparaison des mesures harmoniques en deux points correspondants de deux domaines deduits l'un de l'autre par une transformation analytique uniforme).

Applications: th. des deux constantes de Nevanlinna–Ostrowski, th. des trois cercles d'Hadamard, th. de Phragmen–Lindelöf. Principe de la mesure hyperbolique (augmentation des longueurs non euclidiennes dans les transformations analytiques uniformes) contenant le principe de Lindelöf. Application à la representation conforme (th. de Löwner, th. de la derivée angulaire), à la demonstration des th. de Schottky et

Landau, à l'etude du domaine d'intétermination dune fonction en un point.

IV. Principe de Carleman (augmentation de la mesure harmonique par une extension convenable du domaine) fournissant un procede d'approximation de la mes. harmonique; application à l'etude de la convergence, à l'inegalite de Carleman (utilisee par cet auteur dans l'etude des valeurs asymptotiques des f. entieres). Complements au th. des deux constantes, th. nouveaux sur la mesure harmonique; th. sur la deformation dans la rep. conforme (Koebe et Ahlfors). Application à la resolution du probleme de Carleman–Milloux.

V. Ensemble de points (frontieres d'un domaine G de mesure harmonique nulle, recherche des propriétes de ces ensembles E qui sont de mes. harm. nulle indépendamment de G. Constante de Robin et ensembles E' de capacite nulle. Identite des ensembles E, E'. Probleme de Robin et probleme de Fekete. Théorème sur le comportement dune fonction dans le voisinage des points d'un ensemble E. Théorème de Boutroux–H. Cartan–Ahlfors. Propriétés métriques des ensembles de mes. harm. nulle.

VI. Fonctions méromorphes pour $|z| < R$ (R fini ou non), premier théorème sur la fonction caracteristique $T(r)$ de Nevanlinna (croissance, convexité) demontre par la methode de H. Cartan. Relation entre $T(r)$ et l'indicatrice de Shimizu et Ahlfors.

VII. Fonctions méromorphes pour $|z| < 1$ pour lesquelles $T(r)$ est borné; ce sont les quotients de fonctions bornées. Leur représentation par deux produits de Blaschke et une intégrale de Poisson–Stieltjes. Théorèmes de Fatou et des freres Riesz. Application à l'etude de la représentation conforme de la surface universelle de recouvrement d'un domaine simple (schlicht).

VIII. Fonctions méromorphes d'ordre fini. Produits canoniques; théorème sur la factorisation (Hadamard–Borel–Nevanlinna) demontré par la méthode de F. Nevanlinna. Question du genre.

IX. Théorie generale des fonctions meromorphes. Relation fondamentale entre les moyennes $N(r,a)$ et la fonction $T(r)$. (Dans la demonstration, l'approximation de la moyenne logarithmique $m(r, \frac{w}{w'})$ est faite par la méthode des masses de Ahlfors.) Méthode de F. Nevanlinna.

X. Théorème de Borel–Hadamard et ses complements. Relation des défauts de R. Nevanlinna. Cas des fonctions méromorphes pour $|z| < 1$, th. de Frostman sur les defauts des fonctions à caracteristique $T(r)$ non bornée. Complement au th. de Valiron–Ahlfors sur la limite de $N(r,a) : T(r)$. Théorème sur les racines multiples et th. de Picard sur l'uniformisation des fonctions de genre superieur à 1.

XI. Etude des surfaces S; singularités transcendantes directes et indirectes (Boutroux-Iversen). Théorèmes de Iversen et de Gross sur les surfaces du type parabolique. Surfaces dont les singularités se projettent en p points (Nevanlinna–Elfving), cas où il n'y a qu'un nombre fini de singularités. Application au probleme du defaut. Théorème de Denjoy–Carleman–Ahlfors sur les singularités directes des fonctions d'ordre fini.

XII. Ramxification de S, sa relation avec le défaut total. Notion de ramification totale. Condition suffisante pour que S soit du type hyperbolique (Nevanlinna) ou du type parabolique (Ahlfors). Critere de Kobayashi, application au cas dun nombre fini de points de ramification.

XIII. Exposé detaille de la théorie des surfaces de recouvrement d'après Ahlfors. *Georges Valiron*

Zbl 17.03602
Haimovici, Mendel; Levi-Civita, Tullio
Sugli spazi metrici a connessione affine.
(On metric spaces with affine connections.) (Italian)
Atti Accad. Naz. Lincei, Rend., VI. Ser. 25, 315-320 (1937).

Der Tangentialraum T einer n-dimensionalen Mannigfaltigkeit im Punkte P soll eine affine Gruppe Γ zulassen, welche ein uneigentliches $(n-2)$-dimensionales Gebilde H im T reproduziert; G sei ihre zentroaffine Untergruppe, $L = 0$ die homogene Gleichung (ersten Grades) des Kegels, der H von P aus projiziert, und zwar in Bezug auf ein n-Bein, welches mit dem Koordinaten-n-Bein nicht identisch zu sein braucht. Der Vergleich der Gleichungen für vektorielle Parallelverschiebung, ausgedrückt einmal in bezug auf G, das andere Mal in der üblichen tensoriellen Darstellung, liefert ein Gleichungssystem S, (für die Konnexionskoeffizienten $\Gamma^{\nu}_{\lambda\mu} = \Gamma^{\nu}_{\mu\lambda}$ und für die in der Zwischenrechnung auftretenden unbekannten Funktionen), aus welchem $\Gamma^{\nu}_{\lambda\mu}$ eliminiert werden können, so daß man ein zweites System S_2 (nur für die obenerwähnten unbekannten Funktionen) bekommt. Wird S_2 nach diesen Funktionen gelöst, und werden diese in S_1 eingesetzt, so bekommt man aus S_1 die $\Gamma^{\nu}_{\lambda\mu}$. Die dadurch bestimmte Konnexion ist affin und metrisch (mit der metrischen Form L), da bei der Parallelverschiebung $dL = L d\psi$ (wo $d\psi$ kein exaktes Differential zu sein braucht). Es folgen Beispiele für $n = 3$. *Vaclav Hlavaty (Princeton)*

Author profiles:

Haimovici, Mendel

Spellings: Haimovici, M. [74] Haimovici, Mendel [19] Haimovici, M.Mendel [1] Haimovici, M.M. [1]
Author-Id: haimovici.mendel
Publications: 95 including 2 Book(s) and 88 Journal Article(s)

MSC 2010

2	01 · History; biography
1	35 · Partial differential equations (PDE)
1	53 · Differential geometry
1	58 · Global analysis, analysis on manifolds

Journals

17	Comptes Rendus de l'Académie des Sciences. Paris
7	Atti della Accademia Nazionale dei Lincei
3	Atti della Accademia Nazionale dei Lincei. Serie Ottava. Rendiconti. Classe di Scienze Fisiche, Matematiche e Naturali
1	Analele Ştiinţifice ale Universităţii Al. I. Cuza din Iaşi. (Serie Nouă.) Secţiunea Ia. Matematică-Informatică
1	Annali di Matematica Pura ed Applicata. Serie Quarta

more ...

Co-Authors

2	Popa, Ionuţ
1	Levi-Civita, Tullio
1	Popa, Emil M.
1	Popa, Eugen

Publication Years

Levi-Civita, Tullio

Spellings: Levi-Civita, T. [224] Levi-Civita, Tullio [55] Levi-Cività, T. [5] Lévi-Civita, T. [2]
Author-Id: levi-civita.tullio
Publications: 286 including 38 Book(s) and 242 Journal Article(s)

MSC 2010

5	53 · Differential geometry
2	44 · Integral transforms, operational calculus
2	70 · Mechanics of particles and systems
2	78 · Optics, electromagnetic theory
2	83 · Relativity and gravitational theory

more ...

Journals

13	Comptes Rendus de l'Académie des Sciences. Paris
9	Atti della Accademia Nazionale dei Lincei
6	Mathematische Annalen
4	American Journal of Mathematics
4	Rendiconti del Circolo Matematico di Palermo

more ...

Co-Authors

11	Amaldi, Ugo
2	Ricci, M. G.
1	Corbino, O. M.
1	Fubini, Guido
1	Haimovici, Mendel

more ...

Publication Years

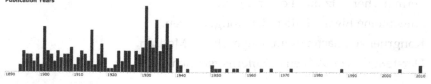

Zbl 20.29201
Hardy, G.H.; Wright, E.M.
An introduction to the theory of numbers.
Oxford: Clarendon Press. xvi, 403 p. bound: 25/- (1938).

Das Buch will kein systematisches Lehrbuch sein, sondern eine Einführung besonderer Art in die Zahlentheorie, richtiger eine Reihe solcher Einführungen geben. Diese Einführungen sind Führungen vergleichbar, die nach und nach fast alle Teile der Zahlentheorie erschließen. Sie erstrecken sich auf die elementare, die analytische und die additive Zahlentheorie, ferner auf die Geometrie der Zahlen, die Theorie der algebraischen Zahlen und die der diophantischen Approximationen. Es ist selbstverständlich, daß bei einem so weit gespannten Rahmen unter gleichzeitiger Beschränkung auf 400 Seiten Vollständigkeit nicht erstrebt werden kann. So wird z.B. auf die Theorie der quadratischen Formen ganz verzichtet; über die Untersuchungen von Hardy und Littlewood zur partito numerorum und daran anschließende Arbeiten wird nur ergebnisweise in einem Anhang berichtet; der Primzahlsatz und der Dirichletsche Satz von der arithmetischen Progression werden nicht bewiesen. Der Leser wird jedoch mit den letztgenannten Sätzen bereits auf den ersten Seiten des Buches bekannt gemacht und auch sonst bei allen behandelten Problemen nach Möglichkeit bis an die neuesten Ergebnisse herangeführt. Kurze Ergänzungen werden vielfach in den Anhängen gegeben, die jedem Kapitel angefügt sind und die im übrigen wertvolle Hinweise auf die Entwicklung der Probleme und auf die Originalliteratur enthalten.

Die Eigenart des Buches liegt in der unkonventionellen Auswahl und Anordnung des Stoffes, in der Mannigfaltigkeit der behandelten Fragen und in ihrer Verbindung untereinander. Die Verff. hatten die Absicht, vor allen Dingen ein interessantes und ungewöhnliches Buch zu schreiben. Das ist ihnen vollkommen gelungen. In außerordentlich klarer Darstellung wird ein Bild der Zahlentheorie entworfen, dessen Schönheit und Lebendigkeit nicht nur im Stoff selbst, sondern ebensosehr in der persönlichen Gestaltungskraft der Verff. begründet liegt, denen die Entwicklung der Zahlentheorie so viel zu verdanken hat.

Das Buch gliedert sich im einzelnen in die folgenden Kapitel:

1. Primzahlen (1. Teil).
2. Primzahlen (2. Teil).
3. Fareyreihen und ein Satz von Minkowski (über Punktgitter).
4. Irrationalzahlen.
5. Kongruenzen und Reste (Grundbegriffe).
6. Fermatscher Satz und Folgerungen.
7. Allgemeine Eigenschaften der Kongruenzen.
8. Kongruenzen nach zusammengesetzten Moduln.
9. Darstellung von Zahlen in Ziffernsystemen.

Author profile:

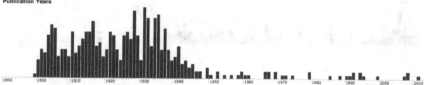

10. Kettenbrüche.
11. Approximation von irrationalen Zahlen durch rationale.
12. Der Fundamentalsatz der Zahlentheorie in den Körpern $k(l)$, $k(i)$ und $k(\rho)$.
13. Einige diophantische Gleichungen.
14. Quadratische Zahlkörper (1. Teil).
15. Quadratische Zahlkörper (2. Teil).
16. Die zahlentheoretischen Funktionen $\varphi(n)$, $\mu(n)$, $d(n)$, $\sigma(n)$, $r(n)$.
17. Erzeugende Funktionen für zahlentheoretische Funktionen.
18. Die Größenordnung von zahlentheoretischen Funktionen.
19. Partitionen.
20. Darstellung von Zahlen durch zwei oder vier Quadrate.
21. Darstellung durch Kuben und höhere Potenzen.
22. Primzahlen (3. Teil).
23. Kroneckers Approximationssatz.
24. Einige weitere Sätze von Minkowski (über Linearformen).

H. Rohrbach (Göttingen)

38

Author profile:

Wright, E.Maitland

Spellings: Wright, E.M. [135] Wright, E.Maitland [20] Wright, E. M. [19] Wright, E. Maitland [9]
Author-Id: wright.e-maitland
Publications: 183 including 8 Book(s) and 174 Journal Article(s)

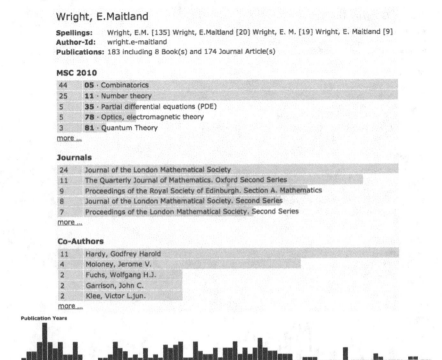

MSC 2010

44	05 · Combinatorics
25	11 · Number theory
5	35 · Partial differential equations (PDE)
5	78 · Optics, electromagnetic theory
3	81 · Quantum Theory

more ...

Journals

24	Journal of the London Mathematical Society
11	The Quarterly Journal of Mathematics. Oxford Second Series
9	Proceedings of the Royal Society of Edinburgh. Section A. Mathematics
8	Journal of the London Mathematical Society. Second Series
7	Proceedings of the London Mathematical Society. Second Series

more ...

Co-Authors

11	Hardy, Godfrey Harold
4	Moloney, Jerome V.
2	Fuchs, Wolfgang H.J.
2	Garrison, John C.
2	Klee, Victor L.jun.

more ...

Publication Years

Zbl 21.25001
van der Waerden, B.L.
Einführung in die algebraische Geometrie.
(Die Grundlehren der mathematischen Wissenschaft in Einzeldarstellungen mit besonderer Berücksichtigung der Anwendungsgebiete. 51) Berlin: Julius Springer. VII, 247 S., 15 Abb. (1939).

Questo volume, dedicato ad un'introduzione alla geometria algebrica, presenta alcune delle ben note caratteristiche delle opere del suo Autore, e precisamente la nitidezza dell'esposizione, la rapidità e compattezza della trattazione, tenuta nei limiti di una severa economià, e la costante aspirazione al rigore ed alla chiarezza nei fondamenti. Non vi si trova invece quel serrato giuoco di concetti astratti, cosi caratteristico della "Moderne Algebra" [Die Grundlehren der mathematischen Wissenschaften in Einzeldarstellungen mit besonderer Berücksichtigung der Anwendungsgebiete Bd. 23. Berlin, J. Springer (1930; JFM 56.0138.01), (1931; JFM 57.0153.03)], che rende quest'ultima di difficile lettura per chi non abbia un'ampia preparazione preliminare. Come afferma l'A., una trattazione di carattere strettamente deduttivo ed astratto della geometria algebrica, che partisse da un corpo qualunque, svolgesse la conseguente teoria delle varietà n-dimensionali, ed ottenesse gli ordinari teoremi della geometria sopra una curva come casi estrema-

Author profile:

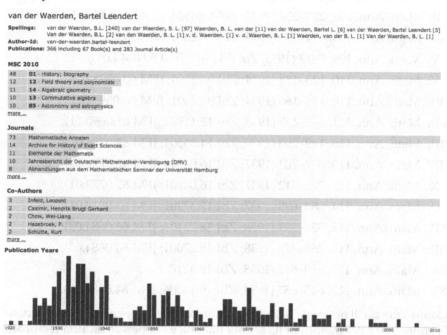

mente particolari di teoremi generali, sarebbe oggidi senz'altro possibile. Ma, opportunamente, trattandosi di un volume introduttivo, l'A. non ha seguito questa strada, procedendo in modo più induttivo e badando sempre a mettere in evidenza il materiale intuitivo, che giustifica l'introduzione di ogni concetto astratto. L.A. opera conseguentemente quasi sempre nel corpo dei numeri complessi e talvolta non rifugge nemmeno da considerazioni di continuità o di teoria delle funzioni. L. A. ha altresi abbandonato i metodi espositivi, di natura aritmetica, basati sulla teoria degli ideali, per sostituirli con i metodi algebrico geometrici, caratteristici della scuola italiana. E' stato cosi possibile accostarsi, sino ad un certo punto, allo sviluppo storico della geometria algebrica. Giova tuttavia dire esplicitamente che il libro si presta piú a dare allo studioso il possesso sicuro e preciso dei fatti esposti che a renderlo padrone dello sviluppo storico dei concetti. Invero l'A. dedica soltanto qua e là poche righe alle notizie storiche; ed è estremamente parco nelle citazioni: ad es., per limitarci al contributo italiano, nessun geometra italiano e citato nella questione della classificazione delle omografie o del principio della conservazione del numero o del teorema dell' $AF + f\Phi$ (secondo l'A. $Af + Bg$). L.A. ha tratto molto materiale dalla sua "Moderne Algebra", alla quale si rimanda per diverse dimostrazioni, e dalla serie delle sue pubblicazioni "Zur algebraischen Geometrie"

40

I: Math. Ann. 108, 113–125 (1933; Zbl 7.07406; JFM 59.0642.11);

II: Math. Ann. 108, 253–259 (1933; Zbl 6.36504; JFM 59.0643.01);

III: Math. Ann. 108, 694–698 (1933; Zbl 7.22604; JFM 59.0643.02);

IV: Math. Ann. 109, 7–12 (1933; Zbl 7.42101; JFM 59.0644.01);

V: Math. Ann. 110, 128–133 (1934; Zbl 9.22504; JFM 60.0595.01);

VI: Math. Ann. 110, 134–160 (1934; Zbl 9.22601; JFM 60.0595.02);

VII: Math. Ann. 111, 432–437 (1935; Zbl 12.11903; JFM 61.0690.11);

VIII: Math. Ann. 113, 199–205 (1936; Zbl 14.36504; JFM 62.0786.01);

IX: Math. Ann. 113, 692–704 (1937; Zbl 16.04004; JFM 62.0772.02);

X: Math. Ann. 113, 705–712 (1937; Zbl 16.04101; JFM 62.0773.01);

XI: Math. Ann. 114, 683–699 (1937; Zbl 17.37301; JFM 63.0612.01);

XII: Math. Ann. 115, 330–332 (1938; Zbl 18.23401; JFM 64.0684.03);

XIII: Math. Ann. 115, 359–378 (1938; Zbl 18.27001; JFM 64.0684.03);

XIV: Math. Ann. 115, 619–642 (1938; Zbl 18.42106);

XV: Math. Ann. 115, 645–655 (1938; Zbl 19.18006; JFM 64.0684.03).

Comunque sia, il notevole libro di vàn der Waerden agevolerà senza dubbio la conoscenza dei metodi della scuola italiana e coopererà ad una reciproca comprensione tra i geometri italiani e gli algebristi tedeschi, assolvendo cosíun compito di grande importanza.

Passiamo ad un rapido esame del contenuto del volume, che consta di 9 eapitoli. Il cap. I e dedicato alla geometria proiettiva degli spazi ad n dimensioni: contiene le generalità sugli spazi proiettivi complessi, lo studio delle trasformazioni proiettive, le coordinate plückeriane, il concetto di geometria astratta. Come esemplificazioni vengono sviluppate le prime proprietà della geometria della retta in S_3; lo studio degli spazi lineari contenuti in una quadrica di S, lo studio della cubica gobba. Il cap. II espone i concetti fondamentali sulle funzioni algebriche. Dopo una breve premessa, che inquadra le funzioni algebriche nel corpo dell'algebra moderna, si tien conto in modo essenziale anche del punto di vista della teoria delle funzioni, sino a pervenire agli sviluppi di Puiseux. Si dà un cenno efficace sulla possibilità, indicata da Ostrowski, di sostituire questi sviluppi con altri di carattere formale e puramente algebrico, meno naturali e semplici di quelli di Puiseux , ma più coerenti alla natura della questione. Si enunciano infine i principali risultati della teoria dell'eliminazione. Il cap. III inizia lo studio delle varietà algebriche, dal caso più semplice delle curve piane: contiene anzitutto una elegante e rigorosa presentazione dei concetti d'irriducibilità, di ordine, di molteplicità in un punto di una curva algebrica e del teorema di Bézout. Seguono le nozioni di curva polare, di classe, di rami di una curva algebrica, una prima classificazione delle singolarità (con la esplicita menzione dei tipi più semplici), le prime formule di Plücker. Queste nozioni si applicano allo studio delle cubiche, per le quali si sviluppa anche una teoria dei gruppi di punti, che anticipa la teoria delle serie lineari sopra una curva qualunque. Il capitolo culmina con lo scioglimento delle singolarità mediante trasformazioni cremoniane e colla dimostrazione dell'invarianza del genere, ottenuta attraverso il concetto di differenziale del corpo di funzioni, definito dalla curva. Se ne deduce l'ultima formula di Plücker. Il cap. IV affronta lo studio generale delle varietà algebriche: si trovano qui esposti, in modo ineccepibile, i delicati concetti d'irriducibilità, di punto generico, nonche la rappresentazione monoidale d Cayley–Halphen. Segue la determinazione delle componenti irriducibili di una qualunque varietà, mediante la teoria dell'eliminazione; ed un breve ma limpido studio delle varietà algebriche dal punto di vista topologico. Il cap. V tratta delle corrispondenze algebriche: illustrato il prineipio di corrispondenza di Chasles sull'esempio dei poligoni di Poncelet, ci si eleva al concetto generale di corrispondenza, attaecando ad esso il principio del computo di costanti, al modo di Schubert. Questo principio viene ampiamente illustrato con numerosi esempi, opportunamente scelti. Il cap. VI e dedicato al concetto di molteplicità. Esso si inizia con la ricerca di ehe cosa avvenga delle soluzioni di un problema geometrico quando si specializzino i dati. Ci si imbatte cosi nel principio della conservazione del numero, per la cui validità, vengono assegnate condizioni sufficienti. Il principio stesso viene illustrato su alcuni esempi, tra cui quello cassico relativo al numero delle proiettività che mutano in se una quaterna di punti di una reltta. Segue un criterio per la molteplicia semplice (non esistenza di una tangente comune a due varietà, intersecantisi) e varie considerazioni sugli spazi tangenti, che culminano nel teorema generale di Bézout. Il cap. VII tratta delle serie lineari (effettive o virtuali) sopra una varietà qualunque posti i concetti fondamentali, tra cui quelli d'immagine proiettiva e di serie lineare semplice o composta, se ne fà, applicazione al caso delle curve, dimostrando, con il metodo indicato da Severi, la possibilità, di trasformare birazionalmente ogni curva in una priva di punti multipli. Il capitolo termina colla dimostrazione dei teoremi di Bertini. Nel cap. VIII viene esposta finalmente la geometria spora una curva, con il metodo algebrico di Brill–Noether. Il teorema fondamentale dell' $AF+B\Phi$ viene enunciato in un'accezione dovuta a Dubreil, secondo il quale tutti i teoremi del tipo dell' $AF+B\Phi$ possono dedursi da un lemma di van der Woude. L'esposizione della teoria procede pol al modo solito, culminando nel teorema di Riemann–Roch. Viene accennata l'estensione allo spazio del teorema dell' $AF+B\Phi$, con applicazioni alle curve gobbe dei primi ordini. Infine, il cap. IX offre un'analisi più approfondita delle singolarità delle curve plane. Vi si introducono i punti multipli infinitamente vicini, il concetto di punti "prossimi" ed il cosidetto "albero delle singolarità" di Enriques.   *Fabio Conforto (Rom)*

Zbl 23.16404
Radon, Johann
Über Tschebyscheff-Netze auf Drehflächen und eine Aufgabe der Variationsrechnung.
Mitt. Math. Ges. Hamburg 8, Tl. 2, 147-151 (1940).

Author profile:

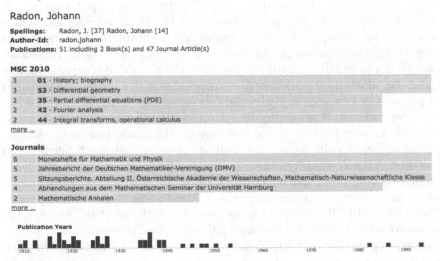

Radon, Johann

Spellings: Radon, J. [37] Radon, Johann [14]
Author-Id: radon.johann
Publications: 51 including 2 Book(s) and 47 Journal Article(s)

MSC 2010

3	01 · History; biography
3	53 · Differential geometry
2	35 · Partial differential equations (PDE)
2	42 · Fourier analysis
2	44 · Integral transforms, operational calculus

more ...

Journals

6	Monatshefte für Mathematik und Physik
5	Jahresbericht der Deutschen Mathematiker-Vereinigung (DMV)
5	Sitzungsberichte. Abteilung II. Österreichische Akademie der Wissenschaften, Mathematisch-Naturwissenschaftliche Klasse
4	Abhandlungen aus dem Mathematischen Seminar der Universität Hamburg
2	Mathematische Annalen

more ...

Publication Years

1910 1920 1930 1940 1950 1960 1970 1980 1990

Verf. bestimmt auf beliebigen Drehflächenzonen äquidistante Netze aus zwei Scharen. von kongruenten, durch Drehung auseinander hervorgehenden Kurven. Ist die Zone der Drehfläche durch

$$x = r(\sigma)\cos\varphi,$$
$$y = r(\sigma)\sin\varphi, \qquad \left(-\pi \le \varphi \le \pi, \quad 0 \le \sigma \le S\right)$$
$$z = z(\sigma)$$

und das Netz durch $u(\sigma,\varphi) = $ const., $v(\sigma,\varphi) = $ const. dargestellt, ist ferner σ die Bogenlänge der Meridiankurve, so ergibt sich

$$u = \alpha\left(\int_0^\sigma \frac{r^2 - \alpha^2 + \beta^2}{rW}d\sigma - \varphi\right); \qquad v = \beta\left(\int_0^\sigma \frac{r^2 + \alpha^2 - \beta^2}{rW}d\sigma - \varphi\right) \qquad (1)$$

wo $W = \sqrt{(\alpha + \beta + r)(\alpha + \beta - r)(\alpha - \beta + r)(-\alpha + \beta + r)}$; $W, \alpha\beta$ können positiv angenommen werden. Durch die Formeln (1) wird die Zone auf ein schlichtes Gebiet der u, v-Ebene ("Netzebene") abgebildet. Läßt man φ variieren, so ist das Bildgebiet der durch

$$0 \leq \frac{u}{\alpha} + \frac{v}{\beta} \leq 2 \int_0^S \frac{r d\sigma}{W} = 2K \tag{2}$$

definierte Streifen der Netzebene. Verf. geht nun umgekehrt von dem durch (2) definierten Streifen der Netzebene aus; die Konstanten α, β, K bestimmen einen "Tchebyscheff Schlauch", der durch (1) auf eine beliebige Drehflächenzone mit bestimmter "Breite" abgebildet wird. Er betrachtet den Sonderfall $\alpha = \beta$, wo die beiden Kurvenscharen des Netzes von zu den Meridiankurven symmetrisch gelegenen Kurven gebildet werden, die Meridiankurven also mit den Netzkurven ein Dreiecksnetz bilden, und zeigt, daß die Forderung eines möglichst großen $z(S)$ bei vorgegebenem $r(0), z(0), r(S)$, die auf eine einfache Variationsaufgabe führt, Drehflächen konstanter negativer Krümmung ergibt.   *O. Volk (Würzburg)*

Zbl 26.08303
Bompiani, Enrico
Invarianti proiettivi e topologici di calotte di superficie e di ipersuperficie
tangenti in un punto. (Projective and topological invariants of surface and
hypersurface caps tangent in a point.) (Italian)
Rend. Mat., Univ. Roma, V. Ser. 2, 261-291 (1941).

Die Arbeit zerfällt in zwei Teile. Im ersten Teil werden die Verfahren entwikkelt, die zur Ableitung der projektiven Ergebnisse erforderlich sind, die Verf. schon in einer vorangehenden Arbeit zusammengestellt hat [Atti Accad. Italia, Rend., VII. 2, 888–895 (1941; Zbl 26.08302)]. Im zweiten Teil beginnt Verf. das Studium der sich in einem Punkt O berührenden Flächenkalotten zweiten Grades des gewöhnlichen Raumes bezüglich der in O regulären Punkttransformationen (topologische Eigenschaften). Die ∞^3 sich in O berührenden Kalotten werden auf die Punkte eines $\mathbb{R}^3$ abgebildet, dann entsprechen den oben genannten Punkttransformationen projektive Transformationen einer G, mit einer invarianten Ebene ω, in der ein nicht zerfallender, invarianter Kegelschnitt Γ liegt; die Punkte von ω bilden die irregulären Kalotten und diejenigen von Γ die parabolischen, irregulären Kalotten ab. Die schon im projektiven Gebiet studierten Begriffe des Kalottenbüschels und Kalottennetzes haben in Wirklichkeit topologischen Charakter. Drei Kalotten eines Büschels besitzen eine Invariante; sie wird im $\mathbb{R}^3$ als das Doppelverhältnis der drei Bildpunkte und des Schnittpunktes ihrer Verbindungsgeraden mit ω gedeutet. Drei Kalotten eines Netzes besitzen zwei Invarianten; sie werden im $\mathbb{R}^3$ gegeben durch zwei der Doppelverhältnisse der drei Bildpunkte und der zwei Schnittpunkte der durch sie bestimmten Ebene mit Γ, wobei diese Doppelverhältnisse auf dem durch die genannten fünf Punkte gehenden Kegelschnitt zu nehmen sind. In den beiden Fällen bestimmt Verf. ein besonders einfaches Modell der betrachteten Konfigurationen, das eine noch unmittelbarere Deutung der Invarianten zuläßt. Die Begriffe werden dann auf Kalotten zweiter Ordnung von Hyperflächen V_n des S_{n+1}, die sich in einem Punkte O berühren, ausgedehnt.   *Piero Buzano (Torino)*

44

Author profile:

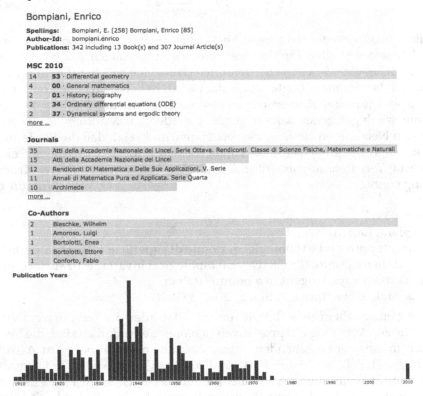
Zbl 28.34101
Hasse, Helmut
Zur arithmetischen Theorie der algebraischen Funktionenkörper.
Jahresber. Dtsch. Math.-Ver. 52, Abt. 1, 1-48 (1942).

Verf. unternimmt es in dieser Arbeit, eine Reihe von Tatsachen aus der Theorie der algebraischen Funktionen einer Unbestimmten, die in den letzten Jabren in vieler Hinsicht fortentwickelt worden ist, zusammenhängend darzustellen. Es handelt sich vor allem um die zahlentheoretische Seite der Theorie und die algebraischen Grundlagen für die Sätze über rationale und ganzzahlige Punktgruppen auf algebraischen Kurven, die entsprechenden Sätze für p-adische Punktgruppen und Punktgruppen mod p von Lutz und Hasse und die algebraische Klassenkörperkonstruktion von Deuring. §1 enthält die Grundbegriffe, algebraische Funktionenkörper K/Ω, Divisoren, rationale Punkte (Primdivisoren ersten Grades), algebraische Punkte (Primdivisoren ersten Grades der durch algebraischen Abschluß des Konstantenkörpers Ω zu $\overline{\Omega}$ entstehenden algebraisch abgeschlossenen Konstantenerweiterung $K : \overline{\Omega}$; §2 die wichtigsten Tatsachen über Divisorenklassen, Riemann-Rochschen Satz und Folgerungen, ferner einen wohl neuen Satz:

Ist M eine primitive Klasse, C eine Klasse mit $\deg C \geq 3g$ (g Geschlecht), so ist $\{MC\} = \{M\}\{C\}$, wo $\{C\}$ den Modul der ganzen Divisoren von C bezeichnet, schließlich die zu einem ganzen Divisor $\mathfrak{m}$ gehörige Erzeugung:

Author profile:

Wenn $1, x_1, \ldots, x_n$ eine Basis von $\mathfrak{m}$ bedeutet, so ist der Integritätsbereich $K^{\mathfrak{m}}$ aller z aus K, deren Nenner in Potenzen von $\mathfrak{m}$ aufgehen, gleich $\Omega[x_1, \ldots, x_n]$, also $K = \Omega(x_1, \ldots, x_n)$. Durch Angabe von $x_1, \ldots, x_n$ und das System (G_0) der algebraischen Gleichungen über Ω zwischen den x_p^* ist K gegeben, daher Erzeugung. §3 führt die homogene Betrachtungsweise ein, wobei der Homogenitätsfaktor, anders als sonst meist üblich, aus K genommen wird, was den Vorteil hat, daß er mit in die Arithmetik von K einbezogen werden kann (würde er als neue Unbestimmte angenommen, so ist das nicht möglich). Ein homogenes Elementsystem $z_0 : z_1 : z_2 : \ldots : z_k$ bestimmt eindeutig ein zu ihm proportionales System teilerfremder ganzer Divisoren $a_0 : a_1 : \ldots : a_k$, die einer Klasse, der Klasse A des Systems $z_0 : z_1 : \ldots : z_k$, angehören. Wenn $z_0 : z_1 : \ldots : z_k$ eine Basis aller ganzen Divisoren von A ist und A hinreichend hohen Grad hat, so wird jeder algebraische Punkt $\overline{\mathfrak{p}}$ von K eindeutig durch die Verhältnisse $z_0(\overline{\mathfrak{p}}) : z_1(\overline{\mathfrak{p}}) : \ldots : z_k(\overline{\mathfrak{p}})$ der z_v mod $\overline{\mathfrak{p}}$ festgelegt. Die $p_v = z(\overline{\mathfrak{p}})$

heißen die homogenen A-Koordinaten von $\overline{\mathfrak{p}}$. Geometrisch bedeutet dies, daß $z_0 : z_1 : \ldots : z_k$ im projektiven k dimensionalen Raum ein singularitätenfreies projektives Modell von K erzeugen. $\Omega(\overline{\mathfrak{p}} = \Omega(\overline{p_1}, \ldots, \overline{p_k}) = \Omega(\overline{p_1} : \ldots : \overline{p_1})$ heißt der Koordinatenkörper von $\overline{\mathfrak{m}}$. Dies läßt sich auf beliebige ganze Divisoren verallgemeinern und führt auf Darstellungen $\overline{\mathfrak{p}} = \frac{(\ldots, p_j z_i - p_i z_j \ldots)}{(z_0, \ldots, z_k)}$ bzw $\mathfrak{a} = \frac{(A_0(z) \ldots A_{ra}(z))}{(z_0, \ldots, z_k)^a}$ der algebraischen Punkte $\overline{\mathfrak{p}}$ bzw. der ganzen Divisoren $\mathfrak{a}$ von K als Quotienten von größten gemeinsamen Teilern von Elementen, d. h. Hauptdivisoren. Dabei sind die $A_i(z)$ Polynome der z_ν mit Koeffizienten aus Ω den A-Koordinaten von $\mathfrak{a}$.

Eine Erzeugung $K = \Omega(z_1, \ldots, z_k)$ kann durch Einführung von Quotienten z_ν/z_0 statt z_ν als homoge Erzeugung $K = \Omega(z_1 : \ldots : z_k)$ mit einem (G_0) entsprechenden homogenen Gleichungssystem (G) zwischen den $z_0, z_1, \ldots, z_k$ aufgefaßt werden. Jedem algebraischen Punkt $\overline{\mathfrak{p}}$ entspricht eine Lösung $p_0 : p_1 : \ldots : p_k$ von (G), es gilt aber auch die Umkehrung.

§4 enthält die Sätze über die Divisorenklassengruppen. Klassen von K heißen rational, solche von $\overline{K\Omega}$ algebraisch, Klassen nullten Grades heißen Nullklassen. Ganze Divisoren g-ten Grades heißen rationale bzw. algebraische Punktgruppen. Ist $\mathfrak{D}$ eine feste Bezugspunktgruppe, so stellen die Brüche $\mathfrak{P} \mathfrak{D} \mathfrak{P}$ beliebige Punktgruppe, alle Nullklassen dar. Für $g > 1$ ist diese Darstellung bei gegebenem $\mathfrak{D}$ nicht immer eindeutig, weil es irreguläre Punktgruppen $\mathfrak{P}$ mit $\dim \mathfrak{P} > 1$ gibt. Die Klasse C heißt $\mathfrak{D}$-regulär, wenn $\dim C\mathfrak{D} = 1$ ist, also C durch einen eindeutig bestimmten Quotienten dargestellt wird. Um die irregulären C zu behandeln, gibt es zwei Mittel: erstens den neuen. Satz: Zu gegebener Anzahl r gibt es eine endliche Menge $\mathfrak{D}_r$ von algebraischen Punktgruppen $\overline{\mathfrak{D}}$, so daß jedes System von r algebraischen Klassen $\overline{C_1}, \ldots, \overline{C_r}$ für mindestens ein $\overline{\mathfrak{D}}$ aus $\mathfrak{D}_r$ $\overline{\mathfrak{D}}$-regulär ist; zweitens die Methode, als Bezugsgruppe die g-te Potenz $\mathfrak{v}^g = \mathfrak{D}$ eines Primdivisors zu nehmen, und C durch $\frac{\mathfrak{P}_\nu}{\mathfrak{D}_\nu}$ mit ganzem $\mathfrak{P}_\nu$ und minimalem ν darzustellen, dieses $\mathfrak{P}_\nu$ ist eindeutig. Der bekannte Zusammenhang dieser Dinge mit der Theorie der Integrale erster Gattung im klassischen Fall wird noch dargestellt. §5 handelt von dem zu K/Ω gehörigen Abelschen Funktionenkörper $\mathcal{K}/\Omega$, der so erhalten wird: zu Ω werden g unabhängige Unbestimmte $x_1, \ldots, x_g$ adjungiert, und in der algebraisch abgeschlossenen Hülle $\overline{\Omega(x_1, \ldots, x_g)}$ wird das Kompositum von g zu $K = \Omega(x, y)$ isomorphen Körpern $K_i = \Omega(x_i, y_i)$ betrachtet, das $g!$ durch $x_i, y_i \mapsto x_{p_i}0, y_{p_i}$ gegebene Automorphismen hat ($p_1, \ldots, p_g$ alle Permutationen von $1, \ldots g$): deren Invariantenkörper ist $\mathcal{K}/\Omega$. Man kann $\mathcal{K}$ auch als Koordinatenkörper einer Punktgruppe der Konstantenerweiterung $K(x_1, \ldots, x_g)\Omega(x_1, \ldots, x_g)$ erhalten, wenn jeder in ihr enthaltene Punkt transzendent, d. h. nicht schon in K/Ω enthalten ist (höchsttranszendente Punktgruppe); jede solche Punktgruppe definiert einen bestimmten Isomorphismus von K/Ω auf einen Teilkörper von $\overline{\Omega(x_1, \ldots, x_g)}/\Omega$. Dies sind die *Darstellungen* von K/Ω.

In §6 werden die Translationen, Spiegelungen und Meromorphismen des Abelschen Funktionenkörpers betrachtet. Ist C eine rationale Nullklasse von K/Ω, so wird durch $\mathfrak{X}' \sim \mathfrak{X}C$ jeder höchsttranszendenten Punktgruppe $\mathfrak{X}$ umkehrbar eindeutig eine andere höchsttranszendente Punktgruppe $\mathfrak{X}'$ zugeordnet, deren Koordinatenkörper $\Omega(\mathfrak{X}')$ mit $\Omega(\mathfrak{X})$ zusammenfällt. Diese Zuordnung $\mathfrak{X} \to \mathfrak{X}'$ und zugleich der entstehende Automorphismus von $\mathcal{K} \cong \Omega(\mathfrak{X}') = \Omega(\mathfrak{X})$ heißt die *Translation* τ_0. Die *Spiegelungen* werden ähnlich erklärt. Es läßt sich nun die Wirkung der Multiplikatoren von K/Ω (vgl. Ref., [J. Reine Angew. Math. 177, 161–191 (1937; Zbl 16.34601)]) auf den Abelschen Funktionenkörper erklären: ein Multiplikator μ ist ein gewisser Endomorphismus $C \to \mu C$ der Massengruppe von K/Ω. Sind $\mathfrak{O}$ und $\mathfrak{U}$ rationale Punktgruppen, $\mathfrak{Y}$ höchsttranszendent, so gibt es in $\Lambda = \Omega(\mathfrak{Y})$ mindestens eine Punktgruppe $\mathfrak{X}$ mit $\frac{\mathfrak{X}}{\mathfrak{O}} \sim \mu\frac{\mathfrak{Y}}{\mathfrak{U}}$. μ heißt regulär, wenn X ebenfalls höchsttranszendent ausfällt, $\mathfrak{X}$ ist dann eindeutig durch $\mathfrak{Y}$, $\mathfrak{O}$, $\mathfrak{U}$ bestimmt, die Darstellung $\mathcal{K} = \Omega(\mathfrak{X})$ des Abelschen Funktionenkörpers ist in der Darstellung $\Lambda = \Omega(\mathfrak{Y})$ enthalten, wodurch, wenn entsprechende Koordinaten von $\mathfrak{X}$ und $\mathfrak{Y}$ einander zugeordnet werden, ein Meromorphismus $\mu_{\mathfrak{O},\mathfrak{U}}$ von Λ/Ω auf $\mathcal{K}/\Omega$ bestimmt ist. $\Lambda = \mathcal{K}_\mu$, der μ- Teilungskörper von $\mathcal{K}$, ist von endlichem Grade über $\mathcal{K}$, dieser Grad heißt die Norm $N(\mu)$ von μ, es gilt dann offenbar $N(\mu_1\mu_2) = N(\mu_1)N(\mu_2)$, was auch für irreguläre μ richtig bleibt, wenn für sie $N(\mu) = 0$ gesetzt wird. Die analytische Theorie liefert im klassischen Fall, daß ein natürlicher Multiplikator n, d.h. die Potenzierung der Klassen mit dem Exponenten n, regulär mit der Norm n^{2g} ist. Für allgemeines Ω ist dies nur im Fall $g = 1$ bewiesen [H. Hasse, Zur Theorie der abstrakten elliptischen Funktionenkörper I, II, III. J. Reine Angew. Math. 175, 55–62 (1936; Zbl 14.14903); ibid., 69-88 (1936; Zbl 14.24901); ibid., 193-208 (1936; Zbl 14.24902); Ref., J. Reine Angew. Math. 177, 161–191 (1937; Zbl 16.34601)], für beliebiges 4g4 steht der Beweis noch aus.

In §7 werden die algebraischen Punkte von $\mathcal{K}$, das sind die Homomorphismen auf algebraische Erweiterungen von Ω, und die Primdivisoren, das sind Bewertungen von $\mathcal{K}/\Omega$, untersucht. Es wird eine homogene Transzendenzbasis $X_0, X_1, \ldots, X_g$ zugrunde geegt. Ein Punkt heißt zu ihr gehörig, wenn $X_0 : X_1 : \ldots : X_g$ durch ihn auf ein bestimmtes Verhältnissystem abgebildet wird, ein Primdivisor, wenn die Reste der X_ν nach ihm ein homogenes System vom Transzendenzgrad g^{-1} über Ω bilden. Ist eine höchsttranszendente Punktgruppe $\mathfrak{X} = \mathfrak{x}_1, \ldots, \mathfrak{x}_g$ von K/Ω, also eine Darstellung $\mathcal{K} = \Omega(\mathfrak{X})$ des Abelschen Funktionenkörpers, gegeben, so entsprechen die algebraischen Punkte von $\mathcal{K}/\Omega$, die zu den symmetrischen Grundfunktionen $X_0, \ldots, X_g$ der g Restpaare $x_0(\mathfrak{x}_i), x_1(\mathfrak{x}_i)$ einer homogenen Transzendenzbasis $x-0, x_1$ von K/Ω gehören, umkehrbar eindeutig den algebraischen Punktgruppen $\mathfrak{P} = \overline{\mathfrak{p}_1} \cdots \overline{\mathfrak{p}_g}$ von K/Ω: um zu dem zugehörigen Punkt von K zu gelangen, hat man im wesentlichen $x_0(\mathfrak{x}_i), x_1(\mathfrak{x}_i)$ nach $\overline{\mathfrak{p}}$ zu reduzieren, Permutationen der $\overline{\mathfrak{p}_i}$ ändern wegen der Symmetrie von $\mathcal{K}/\Omega$ nichts. Wesentlich ist, daß gerade die zu $X_0, \ldots, X_g$ gehörigen Punkte auf diese Weise herauskommen. Der $\overline{\mathfrak{P}}$ zugeordnete Punkt werde mit $\mathfrak{X} \to \overline{\mathfrak{P}}$ bezeichnet.

An Primdivisoren werden nur folgende betrachtet: Jedem Divisor $\mathfrak{a}$ von K entspricht im isomorphen $K\varphi_i$, der durch $z \to z\varphi_i = z(\mathfrak{x}_i)$ gegeben ist, ein Divisor $\mathfrak{a}\varphi_i = \mathfrak{a}(\mathfrak{x}_i)$. Wird für einen Primdivisor $\mathfrak{p}$ $\mathfrak{P}(\mathfrak{x}_i)$ auf die Konstantenerweiterung $K\varphi_i(K\varphi_1 \dots K\varphi_{i-1}K\varphi_{i+1} \dots K\varphi_g)/(K\varphi_1 \dots K\varphi_{i-1}K\varphi_{i+1} \dots K\varphi_g)$ übertragen, so sind die $\mathfrak{p}(\mathfrak{x}_1), \dots \mathfrak{p}(\mathfrak{x}_g)$ ein volles System über $\mathcal{K}/\Omega$ konjugierter Primdivisoren, sie bestimmen also alle den gleichen, mit $\mathfrak{p}(\mathfrak{X})$ bezeichneten Primdivisor von $\mathcal{K}/\Omega$.

Wesentlich ist nun der folgende, von van der Waerden mit algebraisch-geometrischen Methoden bewiesene Satz: Sind zwei Darstellungen $\mathcal{K} = \Omega(\mathfrak{X})$ und $\mathcal{K} = \Omega(\mathfrak{X}')$ von $\mathcal{K}$ gegeben und ist $\frac{\mathfrak{x}}{\mathfrak{O}} \sim \frac{\mathfrak{x}'}{\mathfrak{O}'}$, mit rationalen $\mathfrak{O}, \mathfrak{O}'$, also $\mathfrak{X}$ in $\mathfrak{X}'$ durch eine Translation überführbar, so folgt, wenn $\frac{\mathfrak{P}}{\mathfrak{O}} \sim \frac{\mathfrak{P}'}{\mathfrak{O}'}$ mit regulärem $\overline{\mathfrak{P}'}$ ist, daß der Punkt $\mathfrak{X}' \to \overline{\mathfrak{P}'}$ mit dem Punkt $\mathfrak{X} \to \overline{\mathfrak{P}}$ übereinstimmt und umgekehrt. Es sei nun $\overline{A}$ eine algebraische Nullklasse von K/Ω und X eine feste höchsttranszendente Nullklasse. $\mathfrak{O}$ sei eine rationale Punktgruppe, $\frac{\mathfrak{P}}{\mathfrak{O}}$ in A, $\frac{\mathfrak{x}}{\mathfrak{O}}$ in X. Dann ist nach dem obigen der Punkt $\mathfrak{X} \to \overline{\mathfrak{P}}$ unabhängig von $\mathfrak{O}$ durch A bestimmt. Daher ist auf die Möglichkeit irregulärer $\overline{\mathfrak{P}}$ zu achten, sie wird mittels des Hilfssatzes von §4 behandelt. Auf diese Weise sind den Nullklassen von K/Ω umkehrbar eindeutig Punkte von $\mathcal{K}/\Omega$ zugeordnet, und zwar rationalen Klassen rationale Punkte.

In §8 wird die Zerlegung der Punkte von $\mathcal{K}/\Omega$ im μ-Teilungskörper $\mathcal{K}\mu$ betrachtet, der Multiplikator μ sei dabei regulär und separabel, d. h. $\mathcal{K}\mu/\mathcal{K}$ sei separabel. Es sei $\mathcal{K} = \Omega(\mathfrak{X})$, $\mathcal{K}\mu = \Omega(\mathfrak{Y})$ und $\frac{\mathfrak{x}}{\mathfrak{O}} \sim \mu\frac{\mathfrak{Y}}{\mathfrak{U}}$ mit rationalen $\mathfrak{O}, \mathfrak{U}$. Dann gilt entsprechend dem Satz in §7, und ebenfalls von van der Waerden bewiesen: Ist $\frac{\mathfrak{P}}{\mathfrak{O}} \sim \mu\frac{\overline{\mathfrak{O}}}{\mathfrak{U}}$, $\overline{\mathfrak{O}}$ regulär, dann ist einer der in $\mathfrak{X} \to \overline{\mathfrak{P}}$ enthaltenen Punkte $(\mathfrak{X} \to \overline{\mathfrak{P}})_*$ von $\mathcal{K}\mu$ gleich $\mathfrak{Y} \to \overline{\mathfrak{O}}$ und umgekehrt, entsprechend für reguläres $\overline{\mathfrak{P}}$. Mit seiner Hilfe wird gezeigt: Jede Nullklasse $\overline{D}$ ist in der Form $\overline{D} = \mu\overline{D}_1$ darstellbar. Die Anzahl der Lösungen $\overline{D}_1^*$ dieser Gleichung ist $N(\mu)$. Und hieraus wird wie im Falle $g = 1$ geschlossen, daß $\mathcal{K}_\mu$ über $\mathcal{K}$ unverzweigt ist für das System der X Punkte von $\mathcal{K}\mu/\Omega$, das mit dem der Y-Punkte zusammenfällt. Ist $\mathfrak{X} \to \overline{\mathfrak{P}}$ ein Punkt von $\mathcal{K}/\Omega$, so zerfällt er in alle die Punkte $\mathfrak{X} \to \overline{\mathfrak{O}}_i$ mit $\frac{\mathfrak{P}}{\mathfrak{O}} \sim \mu\frac{\overline{\mathfrak{O}}_i}{\mathfrak{U}}$, es sind also die Klassen $\overline{D}_i$ von $\frac{\overline{\mathfrak{O}}_i}{\mathfrak{U}}$ gerade die Lösungen von $\mu\overline{D}_i = \overline{D}$, $\overline{D}$ Klasse von $\frac{\mathfrak{P}}{\mathfrak{O}}$.

In §9 wird die A. Weilsche Theorie der Distributionen für Funktionenkörper einer Unbestimmten neu entwickelt, wobei an Stelle von Idealen mit Divisoren gearbeitet wird.

Max Deuring (Posen)

Zbl 28.07202
Julia, Gaston
Complex analysis and operator theory of the Hilbert space.
Abh. Preuß. Akad. Wiss., Math.-Naturw. Kl. 1943, No.11, 1-15 (1943).

Verf. hatte sich die Aufgabe gesetzt, die Existenzbereiche der nichtbeschränkten linearen Operatoren des Hilbertschen Raumes zu kennzeichnen und die Operatoren in ihrem ganzen Existenzbereich durch einfache analytische Hilfsmittel darzustellen, immer dann, wenn die Darstellung durch Matrizes versagt. Seine diesbezüglichen Einzelergebnisse hat er in einer Reihe von Noten in den Comptes Rendus und in J. Math. pures appl. veröffentlicht (vgl. [C. R. Acad. Sci., Paris 212, 733–736 (1941; Zbl 25.06501); C. R. Acad. Sci., Paris 212, 829–831 (1941; Zbl 25.06502); C. R. Acad. Sci., Paris 212, 1059–1062 (1941; Zbl 25.18702); C. R. Acad. Sci., Paris 213, 5–9 (1941; Zbl 25.34101); C. R. Acad. Sci., Paris 213, 297–300 (1941; Zbl 25.41301); J. Math. Pures Appl., IX. Sér. 20, 347–362 (1941; Zbl 26.13003); C. R. Acad. Sci., Paris 213, 465–469 (1941; Zbl 216.23301); C. R. Acad. Sci., Paris 214, 591–593 (1942; Zbl 27.32102); C. R. Acad. Sci., Paris 214, 709–710 (1942; Zbl 28.07301)]).

Author profile:

Julia, Gaston

Spellings: Julia, G. [142] Julia, Gaston [96]
Author-Id: julia.gaston
Publications: 238 including 45 Book(s) and 187 Journal Article(s)

MSC 2010

11	11 · Number theory
11	30 · Functions of a complex variable
1	47 · Operator theory

Journals

115	Comptes Rendus de l'Académie des Sciences. Paris
6	Acta Mathematica
4	Journal de Mathématiques Pures et Appliquées. Neuvième Série
2	Bulletin de la Société Mathématique de France
2	Commentarii Mathematici Helvetici

more ...

Co-Authors

5	Beghin, Henri
1	Julia, Roger
1	Picard, Emile

Publication Years

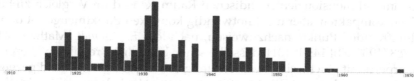

Hier wird über diese Ergebnisse berichtet und auf gewisse Analogien zur Funktionentheorie hingewiesen.

In der Reihenentwicklung

$$Au = \sum_k (u, e_k) f_k \qquad (f_k = Ae_k), \tag{1}$$

wo A ein abgeschlossener Operator mit dem dichten Definitionsbereich D_A, $\{e_k\}$ ein vollständiges orthonormales System in D_A und u ein solches Element aus D_A ist, für welches diese Reihe konvergiert, erblickt Verf. das Analogon der Potenzreihenentwicklung einer analytischen Funktion. Die Menge d_A dieser Elemente u entspricht dem Konvergenzkreis; es ist $d_A \subseteq D_A$, aber im allgemeinen $d_A \neq D_A$ enthält insbesondere die Menge M der endlichen Linearkombinationen der e_k, diese Menge M spielt die Rolle des Inneren des Konvergenzkreises. Für beschränkte Operatoren A, und nur für diese, fällt d_A mit dem ganzen Raum zusammen; diese Operatoren spielen also in dieser Hinsicht die gleiche Rolle wie die ganzen Funktionen in der Funktionentheorie. Ist $dA \neq D_A$, so kann man nach dem Muster der analytischen Fortsetzung verfahren, indem man ein vollständiges orthonormales System in M (d. h. im Inneren von d_A) wählt und den entsprechenden Bereich d'_A, der gegebenenfalls über d_A hinausreicht, betrachtet. Ist das Orthogonalsystem $\{e_k\}$ passend gewählt und werden die Glieder der Summe in der rechten Seite von (1) geeignet gruppiert, so kann man erreichen, daß (1) für jedes $u \in D_A$ gilt; dies entspricht der sog. Überkonvergenz von Potenzreihen.

Für nichtabgeschlossene A gilt die Darstellung (1) im allgemeinen nur für $u \in M$; will man ihre Gültigkeit für ganz d_A bewahren, so muß man auf die Eindeutigkeit des Operators A verzichten. Diese Operatoren spielen also die Rolle der mehrdeutigen Funktionen, während die abgeschlossenen Operatoren die Rolle der eindeutigen Funktionen gespielt haben. Es wird das Problem aufgeworfen, diese mehrdeutigen Operatoren zu "uniformisieren".

Béla de Szökefalvi-Nagy (Szeged)

Zbl 149.19101

Dinghas, Alexander; Schmidt, Erhard
Einfacher Beweis der isoperimetrischen Eigenschaft der Kugel im n-dimensionalen Euklidischen Raum.
Abh. Preuss. Akad. Wiss., Math.-Naturw. Kl. 1943, No.7, 18 S. (1944).

Den Verff. gelingt es zum ersten Male, die isoperimetrische Eigenschaft der Kugel im n-dimensionalen euklidischen Raum $\mathbb{R}^n$ und im Vergleich zu beliebigen kompakten, aber nicht notwendig konvexen Punktmengen $\mathfrak{K}$ ohne "volumfremde" Punkte nachzuweisen, nachdem E. *Schmidt* [Math. Z. 44, 689–788 (1939; Zbl 149.19101)] dasselbe schon bezüglich "regulären" Vergleichskörpern mit stückweise zweimal stetig differenzierbarem Rand gezeigt hatte und nachdem A. *Dinghas* [Sitzungsber. Akad. Wiss. Wien, Math.-Naturw. Kl., Abt. IIa 149, 399–432 (1940; Zbl 24.28401; JFM 66.0910.01)] die letztere Vergleichsklasse noch wesentlich erweitert hatte. Dabei wird nach E. Schmidt unter einem volumfremden Punkt P einer meßbaren Menge $\mathfrak{M}$ des $\mathbb{R}^n$ ein solcher Punkt verstanden, daß der Durchschnitt von mit einer gewissen Vollkugel um P das Maß 0 besitzt. Außerdem wird der Begriff der Minkowskischen Relativoberfläche O zugrunde gelegt, welche durch $O(\mathfrak{K} = \lim_{h\to0} \frac{V(\mathfrak{K}_h)-V(\mathfrak{K})}{h}$ (V = Volumen, $\mathfrak{K}_h$ = Parallelkörper von $\mathfrak{K}$ vom Abstand h) definiert ist. Die von den Verff. benutzte Beweismethode stellt eine Verallgemeinerung des klassischen Induktionsschlusses von H. Kneser und W. Süss [vgl. *T. Bonnesen, W. Fenchel*, Theorie der konvexen Körper. Erg. Math. u. ihrer Grenzgebiete 3, No. 1, Berlin: Julius Springer. VII, 164 S., 8 Fig. (1934; Zbl 8.07708; JFM 60.0673.01)] zum Beweis der Brunn-Minkowskischen Ungleichung für $\mathfrak{K}$ und eine Kugel dar.

Author profile:

Dinghas, Alexander

Spellings: Dinghas, Alexander [91] Dinghas, A. [64] Dinghas, Alexandre [8]
Author-Id: dinghas.alexander
Publications: 163 including 6 Book(s) and 154 Journal Article(s)

MSC 2010

7	30 · Functions of a complex variable
2	32 · Functions of several complex variables and analytic spaces
2	49 · Calculus of variations and optimal control; optimization
1	11 · Number theory
1	31 · Potential theory

more ...

Journals

40	Mathematische Zeitschrift
10	Mathematische Annalen
9	Comptes Rendus de l'Académie des Sciences. Paris
8	Sitzungsberichte der Preussischen Akademie der Wissenschaften
6	Mathematische Nachrichten

more ...

Co-Authors

2	Schmidt, Erhard

Publication Years

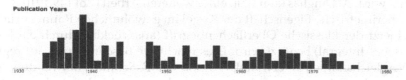

52

Author profile:

Schmidt, Erhard

Spellings: Schmidt, Erhard [43] Schmidt, Erhard. [4]
Author-Id: schmidt.erhard
Publications: 47 including 4 Book(s) and 43 Journal Article(s)

MSC 2010

3	**49** · Calculus of variations and optimal control; optimization
2	**11** · Number theory
2	**45** · Integral equations
2	**55** · Algebraic topology
1	**00** · General mathematics

more ...

Journals

10	Mathematische Annalen
9	Mathematische Zeitschrift
2	Mathematische Nachrichten
2	Zeitschrift für Angewandte Mathematik und Mechanik (ZAMM)
1	Comptes Rendus de l'Académie des Sciences. Paris

more ...

Co-Authors

2	Dinghas, Alexander
1	Carathéodory, Constantin
1	Fuchs, Laszlo
1	Hilbert, David M.

Publication Years

In diesem Zusammenhang sei noch auf einen Isoperimetriebeweis von *F. Baebler* [Arch. Math. 8, 52–65 (1957; Zbl 105.16203)] mittels der Steinerschen Symmetrisierung verwiesen, wo bezüglich $\Re$ dieselben Voraussetzungen wie in der Dinghas–Schmidtschen Arbeit gemacht werden und wo ebenfalls wie dort die Kugel auch als einzige Lösung des isoperimetrischen Problems erkannt wird. A. Dinghas kommt in einer weiteren Arbeit [Zbl 149.19102] auf die isoperimetrische Eigenschaft der Kugel im gewöhnlichen Raum zurück, wobei jetzt der klassische Oberflächenbegriff (ausdrückbar durch ein Riemannsches Integral) benützt wird. Dies macht die Beschränkung auf reguläre Vergleichskörper im erwähnten Sinne von E. Schmidt notwendig; der Nachweis der Extremaleigenschaft der Kugel verläuft hierbei analog zu der zitierten Arbeit von Dinghas aus dem Jahre 1940. *Kurt Leichtweiß*

Zbl 61.41106
Eckmann, Beno
Harmonische Funktionen und Randwertaufgaben in einem Komplex.
Comment. Math. Helv. 17, 240-255 (1945).

Auf einer Teilmenge ≠ 0 von Ecken eines Streckenkomplexes K sei eine reelle Funktion f erklärt; die Definition von f soll auf die übrigen Ecken von K so ausgedehnt werden, daß für jede dieser restlichen Ecken gilt: Der

Author profile:

Wert von f an einer Ecke ist gleich dem arithmetischen Mittel an allen jenen Ecken, die mit der Ecke durch eine Strecke verbunden sind. Das Problem läßt sich mit Hilfe der Begriffe Kette mit reellen Koeffizienten, Rand, Korand, Skalarprodukt von Ketten formulieren; die Lösung wird bewerkstelligt vermöge Orthogonalprojektion im Vektorraum der Ketten, und von hier aus ergeben sich leicht die Bedingungen für Existenz und Eindeutigkeit. Die angegebene Formulierung und Lösung lassen sich auf höhere Dimensionen verallgemeinern. Die Projektionsmethode wird auch auf das Problem der Stromverteilung in einem Leitungsnetz angewendet.   *J. Teixidor*

Zbl 63.06712
Santalo, Luis Antonio
On the convex bodies of constant width in E_n.
Port. Math. 5, 195-201 (1946).

This is a frequently cited integral-geometric study of convex hypersurfaces of constant width in euclidean n-space, providing a good example of the author's excellent and motivating style of writing. It had been published shortly after the end of WW2, when Zentralblatt was in bad shape. Hence in spite of the good quality of the paper so far it had been mentioned in Zentralblatt by title only.

54

Author profile:

Santalo, Luis Antonio

Spellings: Santalo, L.A. [105] Santaló, L.A. [34] Santaló, L. A. [20] Santalo, Luis A. [14] Santaló, Luis A. [8] Santalo Sors, Luis A. [3] Santalo, Luis Antonio [2] Santalo Sors, L. A. [1] Santaló, L.A. [1] Santaló, Luis Antonio [1] Santaló, Luis [1] Santaló, Lius A. [1] Santaló Sors, L. A. [1]

Author-Id: santalo.luis-antonio

Publications: 192 including 16 Book(s) and 164 Journal Article(s)

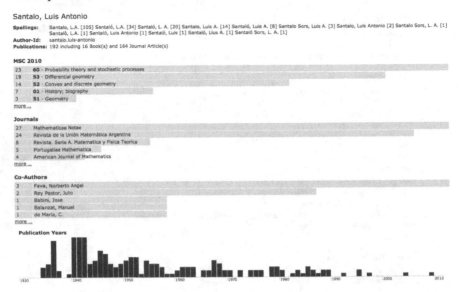

MSC 2010	
23	**60** · Probability theory and stochastic processes
19	**53** · Differential geometry
14	**52** · Convex and discrete geometry
7	**01** · History; biography
3	**51** · Geometry

more ...

Journals	
27	Mathematicae Notae
24	Revista de la Unión Matemática Argentina
8	Revista. Serie A. Matematica y Fisica Teorica
5	Portugaliae Mathematica
4	American Journal of Mathematics

more ...

Co-Authors	
3	Fava, Norberto Angel
2	Rey Pastor, Julio
1	Babini, Jose
1	Balanzat, Manuel
1	de Maria, C.

more ...

Publication Years

1930   1940   1950   1960   1970   1980   1990   2000   2010

The work is based on the famous survey on convexity given by the book of *T. Bonnesen* and *W. Fenchel* [Theorie der konvexen Körper (1934; Zbl 0008.07708), translated into English: Theory of convex bodies. BCS Associates, Moscow, ID (1987; Zbl 628.52001)]. Basic tool is a formula given in Bonnesen-Fenchel, which provides an expression of the volume of an outer parallel hypersurface at distance h to a convex hypersurface in euclidean n-space in terms integrals of the elementary symmetric functions of the principal radii of curvature of the given hypersurface and the distance h. Several interpretations of this formula are presented and extensions to inner parallel hypersurfaces are investigated. In the case of a convex hypersurface bounding a convex body of constant width a the inner parallel hypersurface at distance a coincides with the given hypersurface. This leads to a series of relations between the integral invariants mentioned above. Some of them can be concluded from others. But, as the author shows, more or less half of them are independent. In the case of a planar curve of constant width a we get the well-known relation $L = \pi \cdot a$, L denoting the length of the curve. *Bernd Wegner (Berlin) (2011)*

Zbl 38.29201
Kolmogorov, A.N.
Statistische Theorie der Schwingungen mit kontinuierlichen Spektren.
(Statistical theory of vibrations with continuous spectra.)
Akad. Nauk SSSR, Jubil. Sbornik 1, 242-252 (1947).

Es seien $\xi_r(t)$ $(r = 1, 2, \ldots, n)$ komplexwertige Funktionen des reellen Argumentes t, das im folgenden als Zeit gedeutet werden kann. Wenn die gegebenen Funktionen periodisch oder allgemeiner fastperiodisch sind, können sie in der Form

$$\xi_r(t) \sim \sum_r a_r e^{i\lambda_r t} \quad (r = 1, 2, \ldots, n) \tag{1}$$

dargestellt werden. Es erscheint dann als eine sinngemäße Verallgemeinerung des obigen Ansatzes, einen Grenzübergang von den diskreten zu kontinuierlich veränderlichen Frequenzen vorzunehmen, d. h. die $\xi_r(t)$ als trigonometrische Integrale

$$\xi_r(t) \sim \int_{-\infty}^{\infty} \varphi(\lambda) e^{-i\lambda t} \, d\lambda \quad (r = 1, 2, \ldots, n) \tag{2}$$

darzustellen, oder noch allgemeiner, die beiden erwähnten Fälle in

$$\xi_r(t) \sim \int_{-\infty}^{\infty} e^{i\lambda t} \, d\Phi_r(\lambda) \quad (r = 1, 2, \ldots, n) \tag{3}$$

zusammenzufassen.

Man kann aber bekanntlich Schwingungen mit kontinuierlichen Spektren mit den Mitteln der klassischen Analysis nur als gedämpft darstellen. Es erwies sich aber mehrfach in den physikalischen Anwendungen als notwendig, auch urgedämpfte Schwingungen in der Form (2) und (3) darzustellen, und es wurden tatsächlich Untersuchungen in dieser Richtung von mehreren Autoren unternommen. Wenn auch nicht selten ohne eine strenge mathematische Begründung, so ergaben sich doch mehrfach wertvolle Resultate.

Die mit der Spektraltheorie der Funktionen $\xi_r(t)$ sich, beschäftigenden fachmathematischen Untersuchungen, für die das notwendige Rüstzeug von der Spektraltheorie der Operatoren geliefert wurde, orientierten sich gleich am Anfang an einer strengen Begründung der betreffenden physikalischen Theorien. Aber zu einer endgültigen Lösung der Frage fehlte noch ein entscheidender Schritt, der 1934 von *A. Khintchine* getan wurde [Math. Ann. 109, 604–615 (1934; Zbl 8.36806)]: die Funktionen $\xi_r(t)$ werden in seiner Auffassung als stochastische Variablen in dem üblichen wahrscheinlichkeitstheoretischen Sinne gedeutet, und so ergab sich die Möglichkeit, den analytischen Apparat der Theorie der stochastishen Prozesse in Gang zu setzen. In der Tat: man betrachtet die Gesamtheit der Funktionen $\xi_r(t)$ als einen n-dimensionalen stochastischen Prozeß, und es erscheint als einzige Voraussetzung für die folgenden Überlegungen nur dessen Stationärität im weitesten Sinne [für die Definitionen vgl. *A. Kolmogoroff*, Grundbegriffe der Wahrscheinlichkeitsrechnung. Ergebnisse der Math. und ihrer Grenzgebiete. 2, No. 3. Berlin: Julius Springer (1933; Zbl 7.21601)]. Ziel der vorliegen der Arbeit ist, einige – von sowjetischen wie auch von ausländischen Autoren stammende – Resultate darzulegen, die den Khintchineschen Grundgedanken weiterbilden.

56

Author profile:

Kolmogorov, Andrey Nikolaievich

Spellings: Kolmogorov, A.N. [225] Kolmogoroff, A. [112] Kolmogorov, A. [19] Kolmogoroff, A.N. [11] Kolmogorov, A. N. [9] Kolmogoroff, A. N. [4] Kolmogoroff, Andre [3] Kolmogorov, Andrey Nikolaievich [1] Kolmogorov, Andrey N. [1] Kolmogorov, Andrej N. [1] Kolmogoroff, Andrej [1] Kolmogoroff, A.A. [1] Kolmogoroff [1]
Author-Id: kolmogorov.andrey-nikolaievich
Publications: 389 including 59 Book(s) and 318 Journal Article(s)

MSC 2010

101	**01** · History; biography
30	**60** · Probability theory and stochastic processes
14	**00** · General mathematics
11	**28** · Measure and integration
11	**46** · Functional analysis
more ...

Journals

51	Uspekhi Matematicheskikh Nauk
37	Russian Mathematical Surveys
25	Comptes Rendus (Doklady) de l'Académie des Sciences de l'URSS, Nouvelle Série
20	Comptes Rendus de l'Académie des Sciences, Paris
15	Mathematische Annalen
more ...

Co-Authors

27	Aleksandrov, P. S.
16	Gnedenko, Boris Vladimirovic
15	Fomin, S.V.
12	Gelfand, Israel M.
12	Oleĭnik, Olga Arsenievna
more ...

Publication Years

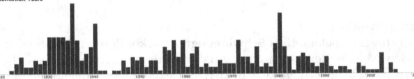

Nach einer Formulierung des Khintchineschen Hauptsatzes nach *H. Cramér* [Ann. Math. (2) 41, 215–230 (1940; Zbl 23.05802)] folgt eine interessante, das Wesentliche des Beweisganges berührende Plausibilitätsbetrachtung. Zur Herstellung der in Formel (3) erscheinenden Funktionen werden die $\xi_r(t)$ als Elemente eines Hilbertschen Raumes betrachtet, in dem die unitäre Metrik durch $(\xi, \eta) = M(\xi\overline{\eta})$ [$M(F)$ Erwartungswert von F] definiert wird. Wie leicht ersichtlich, kann man in diesem Raum durch die Relationen $U^\tau \xi_r(t + \tau)$ $(r = 1, 2, \ldots, n)$ eine einparametrige unitäre Transformationsgruppe erzeugen, die nach *M. H. Stone* [Proceedings USA Academy 16, 172–175 (1930; JFM 56.0357.01)] eine Darstellung der Form $U^\tau = \int_{-\infty}^{\infty} e^{i\lambda\tau} \, dE_\lambda$ zuläßt. [E_λ Zerlegung der Einheit.]
Setzen wir $\Phi_r(\lambda) = E_\lambda \xi_r(0)$ $(r = 1, 2, \ldots, n)$, so erhalten wir nach einer sinngemäßen Deutung des Integralzeichens (3). Es wird dann gezeigt, daß trotz der vielleicht abtrakt erscheinenden Herstellung der Funktionen $\Phi_r(\lambda)$ die Spektralkomponenten

$$\xi_r(t, \Delta_\lambda) = \int_{\Delta_\lambda} e^{i\lambda t} \, d\Phi(\lambda) \quad (r = 1, 2, \ldots, n)$$

durch einen passend gewählten "Filter" für das Experiment zugänglich gemacht werden können.
Besonderes Interesse verdient noch der Fall, daß jede endliche Anzahl von $\xi_r(t)$ eine Gaußsche Verteilung besitzt. In diesem Fall sind die $\Phi_r(\lambda)$ nichtdifferenzierbar, und so haben wir es in diesem für die Anwendungen typis-

chsten Fall mit einem Eindringen der nichtdifferenzierbaren Funktionen in die mathematische Physik zu tun.

Endlich werden noch Arbeiten von *H. Wold* [A study in the analysis of stationary time series. Uppsala (1938; Zbl 19.35602)], *V. N. Zasuchin* [Doklady Akad. Nauk SSSR, n. S. 33, 435–437 (1941; Zbl 0063.08391)] und *E. Sluckij* (Slutzky) [vgl. Voprosy konjunktury 3.I, 34–64 (1927); erweiterte englische Übersetzung in Econometrica, Menasha 5, 105–146 (1937; JFM 63.1131.03))] besprochen, denen auch bei den außerphysikalischen Anwendungen (z. B. Meteorologie, Ökonomie) eine Rolle zukommt.

Auf viele Einzelheiten konnte wegen der Kürze nicht eingegangen werden; für diese muß auf die Originalarbeit verwiesen werden.

Lajos Pukánszky (Budapest)

Zbl 37.35302
Yosida, Kôsaku
On the differentiability and the representation of one-parameter semi-group of linear operators.
J. Math. Soc. Japan 1, No. 1, 15-21 (1948).

Soit $t \to U_t$ une représentation continue bornée ($\|U_t\| \le 1$) du semi-groupe additif des nombres réels positifs dans l'algèbre (munie de la topologie de convergence simple) des opérateurs linéaires continus d'un espace de Banach complexe E. L'A. montre que le sous-espace vectoriel D de E sur lequel existe la dérivée faible $Ax =\underset{h\downarrow 0}{\longrightarrow}$ lim.faible $h^{-1}(U_h - 1)x$ est dense dans E, que $x \to Ax$ est un opérateur fermé et que $h^{-1}(U_{s+h} - U_s)x$ converge dans de plus fortement quand $h \downarrow 0$ vers $AU_s x = U_s Ax$. En introduisant les intégrales $\int_0^\infty \delta e^{-\delta s} x\, ds$ qui représentent des éléments de D et qui convergent fortement vers x quand $\delta \to \infty$, l'A. montre que, si I est l'opérateur unité, les opérateurs $A - nI$ ($n > 0$) sont inversibles et que si $B_n = n(A - nI)^{-1}$ alors $\|B_n\| \le 1$ et $-n(I + nB_n)$ converge simplement vers A sur D quand $n \to \infty$. Réciproquement si A est un opérateur défini sur un sous-espace dense de E, pour lequel les B_n existent et satisfont aux propriétés précédentes alors A est dérivée d'un semi-groupe $t \to U_t$. Application: théorème de Stone (E espace de Hilbert, U_t unitaire).

J. Riss (Saigon)

Zbl 61.09101
Takagi, Teiji
Zur Axiomatik der ganzen und der reellen Zahlen.
(On the axiomatic of integer and real numbers.) (Japanese)
Proc. Japan Acad. 21 (1945), 111-113 (1949).

Verf. legt die Totalität N der ganzen positiven und negativen Zahlen (einschließlich Null) zugrunde.

Axiom 1: N läßt eine nicht-identische, eineindeutige Abbildung in sich zu: $x \leftrightarrow \varphi(x)$. Verf. schreibt x^+ und x^- für $\varphi(x)$ bzw. $\varphi^{-1}(x)$, und $M^\pm (M \subset N)$ für die Menge $\{x^\pm ; x \in M\}$.

Author profile:

Yosida, Kosaku

Spellings: Yosida, Kosaku [69] Yosida, K. [64] Yosida, Kôsaku [31] Yosida, Kōsaku [1]
Author-Id: yosida.kosaku
Publications: 165 including 17 Book(s) and 142 Journal Article(s)

MSC 2010

8	47 · Operator theory
7	34 · Ordinary differential equations (ODE)
6	46 · Functional analysis
5	44 · Integral transforms, operational calculus
3	01 · History; biography

more ...

Journals

7	Proceedings of the Japan Academy. Series A
5	Journal of the Mathematical Society of Japan
4	Nagoya Mathematical Journal
2	Annals of Mathematics. Second Series
2	Journal of the Faculty of Science. Section I A

more ...

Co-Authors

7	Kakutani, Shizuo
4	Kakutani, Shin-ichiro
3	Fukamiya, Masanori
2	Mimura, Yukio
1	Chikazumi, Shinpei

more ...

Publication Years

Axiom 2: Wenn $M \subset N$ und $M = M^+$ ist, so fällt M mit N zusammen (Prinzip der vollständigen Induktion). Nun heiße eine Untermenge M von N eine Progression, wenn für ein Element x von M immer x^+ in M enthalten ist. Es gibt alsdann zwei Fälle:

1. Jede Progression fällt mit N zusammen. In diesem Falle ist N ein Zyklus mit endlich vielen Elementen z. Es gibt eine Progression, welche ein echter Teil von N ist. Dann ist N mit dem System der gewöhnlichen ganzen Zahlen isomorph. Ganz kurz spricht Verf. über die reellen Zahlen.

Axiom 1: Die Menge N der reellen Zahlen ist ein Linearkontinuum, d. h. N ist geordnet, im Dedekindschen Sinne stetig und offen.

Axiom 2: Jedes Linearkontinuum, welches in N enthalten ist, ist mit N ähnlich in bezug auf die Ordnung in N. Auf Grund von Axiom 1 nehmen wir in N eine beiderseits unendliche Folge an, deren Elemente wir ordnungsgemäß mit den ganzen Zahlen bezeichnen. Durch fortgesetzte Zweiteilung erhalten wir somit für die Elemente von N eine Darstellung in der Form dyadischer Brüche.

Zyoiti Suetuna

Author profile:

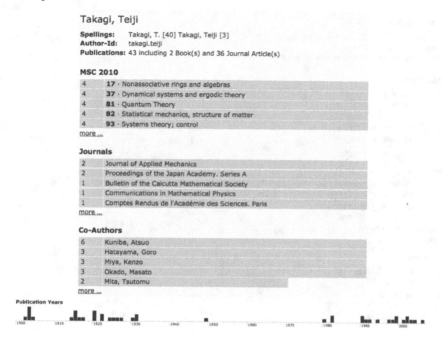

Zbl 39.04701
Sierpiński, Wacław
Les ensembles projectifs et analytiques.
Mem. Sci. Math. 112. Paris: Gauthier-Villars. 80 p. (1950).

Un exposé bien réussi donnant, de la première main, presque tous les résultats et problèmes, qu'on avait trouvés et publiés dans la théorie des ensembles analytiques et projectifs (inventés par Souslin en 1917 et Lusin en 1925 respectivement). Plus de 44 auteurs et tant de périodiques [presque tous les volumes des Fundamenta Mathematicae (pourquoi sans indication des années!)] y sont cités. Un heureux choix de démonstrations complètes et de renvois respectivement. Les pages 28–32 constituent un touchant exposé de la naissance des ensembles analytiques car "Par hasard j'etais present au moment où Michel Souslin communique à M. Lusin sar remarque [relative à la faussete de $\bigcap_n \text{proj} E_n = \text{proj} \bigcap E_n$ si $E_1 \geq E_2 \geq \ldots$ et dont se servait Lebesgue en "prouvant" la fausse proposition que la projection de tout ensemble borelien serait un ensemble borelien] et lui donna le manuscrit de son Premier travail" (pp. 28/29). Les ensembles projectifs sont définis à partir des ensembles ouverts cartésiens par des complementations et projections prises un nombre fini de fois. La méthode des ensembles universels (Lusin), et des cribles très variés (cf. no. 31–38) sont bien mises à point. L'ouvrage se compose de deux parties: les ensembles projectifs (cf. no. 1–16, pp. 1–27) et les ensembles analytiques (no. 17–43, pp. 28–78).   *Duro Kurepa (Zagreb)*

60

Author profile:

Sierpiński, Wacław

Spellings: Sierpiński, Wacław [479] Sierpiński, W. [458] Sierpiński, W. [13] Sierpinski, W. [7] Sierpiński, Wacław [1]
Author-Id: sierpinski.waclaw
Publications: 958 including 54 Book(s) and 892 Journal Article(s)

MSC 2010

20	11 · Number theory
13	03 · Mathematical logic
4	01 · History; biography
4	54 · General topology
1	00 · General mathematics

more ...

Journals

313	Fundamenta Mathematicae
24	Comptes Rendus de l'Académie des Sciences. Paris
23	Wiadomości Matematyczne
15	Elemente der Mathematik
14	Le Matematiche

more ...

Co-Authors

10	Schinzel, Andrzej
7	Lusin, Nicolas
6	Kuratowski, Casimir
6	Szpilrajn, Edward
5	Ruziewicz, St.

more ...

Publication Years

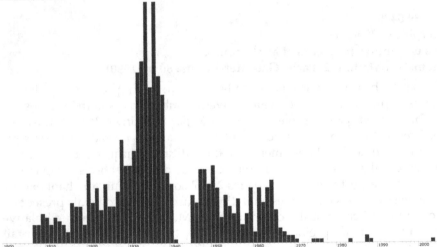

Zbl 42.16202
Haupt, Otto
Über eine Beziehung zwischen Ordnung und Singularitäten.
Math. Nachr. 4, 81-96 (1951).

Le point de départ des considérations de cet article fut la proposition suivante: O désignant un ovale du plan (courbe fermée strictement convexe) possédant en chaque point x un vrai cercle osculateur libre (ou de Menger) $C(x)$ et n'ayant qu'un nombre fini de sommets [points où $C(x)$ est surosculateur], une condition nécessaire et suffisante pour que l'ordre cyclique de O soit $= 4$ est qu'aucun cercle ne touche O en plus de deux points. Cette proposition ressortissant de la géométrie infinitésimale directe (selon Bouligand) est établie comme cas particulier d'une proposition purement topologique, l'intérêt se portant sur les systèmes d'au moins 4 points d'intersection de O et des cercles du plan, et leurs déformations continues. Cadre de l'étude abstraite – $\Re$ désigne un cercle orienté, $\mathfrak{m}$ un ensemble de systèmes finis non ordonnés de points de $\Re$, dits "complexes". L'ordre $\mathfrak{D}(\alpha)$ du complexe $\alpha = (P_1, \ldots, P_r)$ de $\mathfrak{m}$ est défini comme $= r$ et l'ordre $\mathfrak{D}(\mathfrak{m})$ de $\mathfrak{m}$ comme la borne supérieure (éventuellement infinie) de $\mathfrak{D}(\alpha)$. Un complexe α est appelé "maximal" si $\mathfrak{D}(\alpha) = \mathfrak{D}(\mathfrak{m})$. Hypothèses – (H1) $\mathfrak{D}(\mathfrak{m})$ est fini $= h$; la borne inférieure de $\mathfrak{D}(\alpha)$ représentée par $k + 1$ est > 3. (H2) Tout système γ de k points distincts de $\Re$ définit un unique système $\alpha = \alpha(\gamma)$ le contenant. Les hypothèses de continuité (H3) régissent la dépendance entre γ et $\alpha(\gamma)$. La notion essentielle y intervenant est celle de régularité: α est dit "reguliér" si l'ordre des complexes $\alpha' = \alpha(\gamma')$ correspondant aux γ' suffisamment voisins d'un sous-système γ de k points de α, est $\mathfrak{D}(\alpha)$. L'hypothèse de monotonie (H4) régit le déplacement des points mobiles d'un complexe α dont $(k - 1)$ sont fixés. La propriété pour un ovale de ne pas admettre de cercles le touchant en plus de deux points est transposée ainsi: Il existe une constante positive b qui peut être supposée $< \frac{1}{4}$ ayant la propriété suivante: $\alpha = (P_1, \ldots, P_r)$ désignant un complexe, s'il existe des arcs ouverts de $\Re$ disjoints $\Re_1, \ldots, \Re_t$, $\Re_\tau$ ($\tau = 1, 2, \ldots, t$) étant de longueur $< b$ et contenant $q_\tau \geq 2$ des points $P_1, \ldots, P_r$, alors (pour tout τ) $(q_r \leq k) \to \Sigma(q_\tau - 1) \leq k - 1$, (il existe un τ) $(q_r \geq k + 1) \to ((t = 1)\&(q_t = k + 1))$. b est dite "borne de concentration" de $\mathfrak{m}$. Un complexe α comprenant au moins k points situés sur un arc de $\Re$ de longueur $< b$ est appelé "concentré". Aucune hypothèse ne fait intervenir de configurations limites. Malgré la présence de la borne b, l'hypothèse de concentration limitée est topologique, $\Re$ peut être regardé comme un cercle topologique; du à sa compacité, toutes les métrisations de $\Re$ sont équivalentes. Théorème abstrait – Sous les hypothèses (H1), (H2), (H3) et (H4), $\mathfrak{D}(\mathfrak{m}) = k + 1$ dans le cas et seulement dans le cas où $\mathfrak{m}$ est de concentration limitée. Marche de la démonstration – (A) A tout complexe maximal α correspond un complexe concentré $\overline{\alpha}$ du même ordre. (B) Il existe (au moins) un complexe concentré β d'ordre $k + 1$. (C) Tous les complexes concentrés sont d'ordre $\geq k + 1$. Applications du Théorème abstrait. Questions connexes. – Du Théorème abstrait est déduit le théorème "concret" de l'Introduction.

Divers théorèmes analogues dus à Böhmer, Carleman, Mohrmann et l'auteur sont rassembles. Dans chacun, d'une propriété ponctuelle ou locale, est déduite une propriété globale. Ils ne sont pas inclus dans la présente synthèse dont sont déduits deux corollaires; l'un concerne un ovale et les coniques déterminées par cinq quelconques de ses points, l'autre une courbe simple fermée du plan projectif pourvu d'une tangente continue et les droites déterminées par deux de ses points. (Remarque du référent – (H2) entraîne dans les applications euclidiennes ou projectives, une limitation sur $\Re$ lorsque les caractéristiques d'ordre C ne sont pas toujours définies par k points distincts de $\Re$, ainsi dans l'espace euclidien R_3, si les C sont des plans, $\Re$ ne doit pas admettre de droite le rencontrant en trois points distincts.)   *Christian Pauc*

Zbl 46.33103
Riesz, Frédéric; Sz.-Nagy, Béla
Leçons d'analyse fonctionnelle.
Budapest: Académie des Sciences de Hongrie. VIII, 448 p. (1952).

Das Buch zerfällt in zwei Teile, der erste gibt eine Theorie der Ableitung und der Integration und stammt von F. Riesz, der zweite behandelt die Integralgleichungen und die linearen Transformationen und hat B. Sz.-Nagy zum Verf. Das Werk ist aus Vorlesungen beider Verff. entstanden, die hervorragend klare und verständliche Darstellung hält etwa die Mitte zwischen dem Stil einer Vorlesung und einer rein systematischen Entwicklung des Gegenstandes. Es wird weniger größte Allgemeinheit angestrebt, viehmehr dem Gange der historischen Entwicklung folgend ein möglichst vielseitiges Bild der Ideen und Methoden an markanten Beispielen zu geben versucht.

Author profile:

Riesz, Frederic

Spellings: Riesz, F. [83] Riesz, Frederic [14] Riesz, Friedrich [8] Riesz, Fr. [7] Riesz, Frigyes [5] Riesz, Frédéric [5] Riesz, Frederick [4]
Author-Id: riesz.frederic
Publications: 126 including 14 Book(s) and 109 Journal Article(s)

MSC 2010

5	46 · Functional analysis
4	47 · Operator theory
3	26 · Real functions
3	28 · Measure and integration
3	45 · Integral equations

more ...

Journals

11	Comptes Rendus de l'Académie des Sciences. Paris
6	Journal of the London Mathematical Society
5	Acta Mathematica
3	Jahresbericht der Deutschen Mathematiker-Vereinigung (DMV)
3	Mathematische Annalen

more ...

Co-Authors

12	Szokefalvi-Nagy, Bela
3	Lorch, Edgar Raymond
1	Fejer, Lipot
1	Rado, Tibor
1	Riesz, Marcel

more ...

Publication Years

Der erste Teil beginnt mit einem direkten Beweis des Satzes von Lebesgue, daß jede Funktion beschränkter Variation fast überall differenzierbar ist. Es schließen sich an weitere Sätze über Funktionen beschränkter Variation und der Satz von Denjoy-Young-Saks über die vier Derivierten einer beliebigen Funktion. Das erste Kapitel (über die Ableitung) schließt mit der Differentiation und Integration von Intervallfunktionen. Das zweite Kapitel bringt das Lebesguesche Integral, das direkt ohne Einführung des Maßbegriffes

64

erklärt wird: Es wird eingeführt für Treppenfunktionen, dann für die Grenzfunktionen wachsender Folgen von Treppenfunktionen mit beschränkten
Integralen, schließlich für die Differenzen solcher Funktionen. Die damit
erreichte Klasse fast überall erklarter Funktionen heißt die Klasse C_2 der
summierbaren Funktionen. Der Satz von B. Levi sagt dann aus, daß C_2
gegenüber der nochmaligen Bildung von Grenzfunktionen wachsender Folgen mit beschränkten Integralen abgeschlossen ist. Es schließen sich an der
Satz von Lebesgue über gliedweise Integration einer majorisierten Folge
summierbarer Funktionen, die Summierbarkeit zusammengesetzter Funktionen, das Lemma von Fatou, die Ungleichungen von Hölder, Minkowski.
Die meßbaren Funktionen werden als Grenzfunktionen summierbarer Funktionen erklärt, das Maß einer Punktmenge als Integral über die charakter-
istische Funktion. Absolute Stetigkeit als Kennzeichen eines unbestimmten
Integrals, partielle Integration, Substitutionsregel folgen. Ein Abschnitt über
den Raum L^2 bringt den Satz von Riesz–Fischer, die Darstellung linearer
Funktionale auf L2, Orthogonalsysteme, Komplementärzerlegungen von L^2.
Auch für die L^p werden die linearen Funktionale bestimmt. Es folgt die
Integration von Funktionen mehrerer Variablen, der Satz von Fubini, die
Ableitung und Integration von Rechtecksfunktionen. Der nächste Abschnitt
bringt den Nachweis, daß die ursprüngliche Lebesguesche Definition des
Integrals der hier gegebenen äquivalent ist und ergänzt die Integrationsthe-
orie (Sätze von Egoroff, Lusin). Kap. III untersucht die linearen Funktionale
auf dem Raum C der stetigen Funktionen und zeigt, daß das Stieltjesintegral
als lineares Funktional auf C erklärt werden kann. Dieses von F. Riesz stammende grundlegende Ergebnis wird ausführlich dargestellt und die Auffassung des Integrals als lineares Funktional und seine Fortsetzung mittels des
Satzes von Hahn–Banach als allgemeines Prinzip ausdrücklich betont. Der
erste Teil schließt mit der Untersuchung des Integrals von Daniell und dem
Satz von Radon–Nikodym.

Author profile:

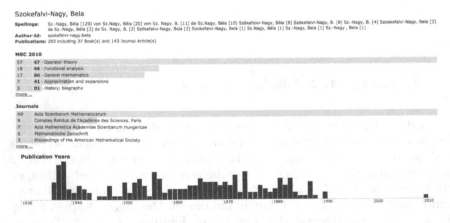

Der zweite Teil beginnt in Kap. IV mit einer Darstellung der Theorie der Integralgleichungen. Volterrasche Integralgleichungen, die Neumannsche Reihe, die Approximation quadratisch integrierbarer Kerne durch Kerne endlichen Ranges bilden den Anfang. Die Alternative von Fredholm wird auf zwei Wegen abgeleitet, einmal nach E. Schmidt durch Approximation durch Kerne endlichen Ranges, dann nach der Methode von F. Riesz, der die Zerlegung von L^2 in zwei komplementare Teilräume bewies, in deren einem $(1 - T)^n f = 0$ gilt, im anderen $f = (1 - T)^n g$ für ein festes geeignetes n, T ein vollstetiger Operator. Auch die Methode der Fredholmschen Determinanten wird kurz auseinandergesetzt. Anwendungen auf das Dirichletsche und das Neumannsche Problem der Potentialtheorie.

Kap. V bringt die axiomatische Einführung des Hilbertschen Raumes, die Theorie der vollstetigen Operatoren in abstrakter Form, zuerst im Hilbertschen Raum, dann die Übertragung auf beliebige Banachsche Räume und als Beispiel zur Theorie der Banachschen Raume die Bestimmung aller linearen Operatoren des Raumes C als Integraltransformationen und die Charakterisierung der vollstetigen unter ihnen (im wesentlichen nach Radon). Kap. VI bringt die Theorie der symmetrischen vollstetigen Operatoren des Hilbertschen Raumes, die Existenz der Eigenwerte und Eigenfunktionen und deren Vollständigkeit nach den Methoden von F. Riesz und O. Kellogg. Diese Theorie wird angewandt auf die symmetrischen und hermiteschen Kerne, der Satz von Mereer wird bewiesen, die schwingende Saite nach Sz.-Nagy behandelt und der Hauptsatz der Theorie der fastperiodischen Funktionen nach H. Weyl und F. Rellich abgeleitet. Kap. VII bringt die Spektralzerlegung für beschrankte symmetrische, unitäre und normale Operatoren nach F. Riesz im wesentlichen, als Beispiele die Fourier-Plancherelsche und die Watsonschen Abbildungen. Kap. VIII enthalt die Spektraltheorie der selbstadjungierten Operatoren, die Fortsetzung symmetrischer Operatoren zu maximalen Operatoren und die Theorie der halbbeschränkten Operatoren nach K. Friedrichs und M. Krejn. Kap. IX enthält den Operatorenkalkül der Funktionen u (AY eines selbstadjungierten Operators A, speziell den Satz, daß jeder lineare abgeschlossene Operator, der mit allen mit A vertauschbaren beschränkten Operatoren vertauschbar ist, eine Funktion von A ist. Es schließt sich eine Darstellung der Hauptergebnisse der Störungstheorie von F. Rellich und Sz.-Nagy an. Kap. X behandelt den Satz von Stone über einparametrige Gruppen unitärer Transformationen sowie den entsprechenden Satz von Sz.-Nagy und Hille über einparametrige Halbgruppen beschränkter selbstadjungierter Operatoren; einige Sätze über Halbgruppen allgemeiner beschränkter Operatoren von Hille u. a., den statistischen Ergodensatz von J. v. Neumann und Verallgemeinerungen davon. Kap. XI bringt kurz die Grundbegriffe der allgemeinen Spektraltheorie in Banachschen Räumen von F. Riesz, Dunford und Lorch, vor allem die Zerlegung eines Operators nach den isolierten Bestandteilen seines Spektrums. Als Anwendung wird der Satz von Wiener über absolut konvergente trigonometrische Reihen bewiesen. Das Buch schließt mit einer kurzen Darstellung der Theorie der Spektralmengen, die kürzlich J. v. Neumann gab.

[Die Verff. haben die Schriftleitung gebeten,. die folgende Ungenauigkeit im Buche richtigzustellen: Auf S. 33 wird behauptet, daß jede Funktion $f(x)$, die zugleich mit $-f(x)$ in die Klasse C gehört, Riemann–integrierbar ist. Das ist so unrichtig, da ja die Funktionen der Klasse C nur bis auf eine Nullmenge definiert sind. Es gilt aber, daß jede solche Funktion $f(x)$ fast uberall gleich einer Riemann–integrierbaren Funktion ist.]   *Gottfried Köthe*

Zbl 50.33502
Köthe, Gottfried
Dualität in der Funktionentheorie.
J. Reine Angew. Math. 191, 30-49 (1953).

Author profile:

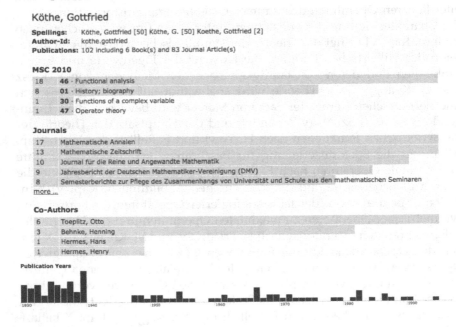

Köthe, Gottfried

Spellings: Köthe, Gottfried [50] Köthe, G. [50] Koethe, Gottfried [2]
Author-Id: kothe.gottfried
Publications: 102 including 6 Book(s) and 83 Journal Article(s)

MSC 2010

18	46 · Functional analysis	
8	01 · History; biography	
1	30 · Functions of a complex variable	
1	47 · Operator theory	

Journals

17	Mathematische Annalen
13	Mathematische Zeitschrift
10	Journal für die Reine und Angewandte Mathematik
9	Jahresbericht der Deutschen Mathematiker-Vereinigung (DMV)
8	Semesterberichte zur Pflege des Zusammenhangs von Universität und Schule aus den mathematischen Seminaren

more ...

Co-Authors

6	Toeplitz, Otto
3	Behnke, Henning
1	Hermes, Hans
1	Hermes, Henry

Publication Years

La théorie des fonctionnelles analytiques de Fantappiè a été rattachée par *J. Sebastião e Silva* [Port. Math. 9, 1–130 (1950; Zbl 0041.43802)] á la théorie générale des espaces vectoriels topologiques, mais sans qu'il parvienne á munir de topologies satisfaisantes les espaces qu il considérait. Cette lacune a été comblée par C. L. de Silva Dias, G. Köthe et A. Grothendieck, qui ont (indépendamment) cu l'idée d'appliquer au problème de Silva la théorie des espaces (F) et (LF) et fait rentrer ainsi la théorie de Fantappiè dans la théorie générale de la dualité entre espaces vectoriels topologiques. Le travail de l'Auteur donne essentiellement la description de cette dualité entre les deux espaces $P(O)$ et $R(\Omega - O)$ définis comme suit: Ω étant la sphère de Riemann,

O un ensemble ouvert dang 0, $P(O)$ est l'espace des fonctions analytiques dans O et nulles au point ∞ si ce point est dans O, avec la topologie de la convergence compacte qui en fait un espace de Montel (et aussi un espace de Fréchet); $R(\Omega - O)$ est l'espace des germes de fonctions analytiques sur $\Omega - O$ limite inductive des espaces de Banach formé des fonctions analytiques dans les voisinages fermés de $\Omega - O$, ayant pour intersection cet ensemble. La dualité est realisée par l'intégrale

$$\langle u, x \rangle = \frac{1}{2\pi i} \oint u(t)x(t)dt$$

étendue á un système convenable de courbes. Ces résultats sont identiques á ceux déja, publiés par C. L. de Silva Dias [Bol. Soc. Mat. Sao Paulo 5, 1–58 (1952; Zbl 49.20001)]. En outre, l'Auteur determine la forme générale d'une application linéaire continue de $P(O_1)$ dann $P(O_2)$, et examine certains types particuliers de telles applications. *Jean Dieudonné*

Zbl 57.15603
Ehresmann, Charles
Extension du calcul des jets aux jets non holonomes.
C. R. Acad. Sci., Paris 239, 1762-1764 (1954).

Soient V_m et V_n, deux varietés de classe $\geq r$, $J^k(V_n, V_m)$ la varieté des jets d'ordre k de V_n on dans V_m, α la projection source de $J^k(V_n, V_m)$ sur V_n. Soit σ un relevèment local de V_n dans $J^k(V_n, V_m)$, tout jet $j^l_x \sigma$ sera un jet non holonome de V_n dans V_m. L'espace de ces jets ist uni sousvarieté de $J^l(V_n(J^k(V_n, V_m))$. On definit encore le prolongement non-holonome général d'ordre l en prenant l fois un prolongement non-holonome d'ordre 1. Les jets nonholonomes formes avec des jets verifiant $j^l_x(j^{k-1} \circ \sigma) = \sigma(x)$ donnent les jets semi-holonomes. Il ist vérifié que les regles de calcul et les principales propriétés des jets établis dans les notes antérieures [C. R. Acad. Sci. Paris 233, 598–600 (1951; Zbl 43.17401); Zbl 46.40703; C. R. Acad. Sci., Paris 234, 587–589 (1952; Zbl 46.40801); C. R. Acad. Sci., Paris 234, 1424–1425 (1952; Zbl 46.40802)] se conservent, mutatis mutandis, pour les jets non-holonomes et semi-holonomes. *Heinrich Guggenheimer (West Hempstead)*

68

Author profile:

Ehresmann, Charles

Spellings: Ehresmann, Charles [114] Ehresmann, C. [21]
Author-Id: ehresmann.charles
Publications: 135 including 17 Book(s) and 99 Journal Article(s)

MSC 2010

30	**18** · Category theory, homological algebra
7	**01** · History; biography
2	**08** · General algebraic systems
2	**53** · Differential geometry
2	**58** · Global analysis, analysis on manifolds

more ...

Journals

53	Comptes Rendus de l'Académie des Sciences. Paris
9	Cahiers de Topologie et Géométrie Différentielle Catégoriques
4	Annales de l'Institut Fourier
2	Annals of Mathematics. Second Series
2	L'Enseignement Mathématique

more ...

Co-Authors

8	Ehresmann, Andree Charles
3	Libermann, Paulette
2	Feldbau, Jacques
1	Calabi, Lorenzo
1	Reeb, Georges H.

more ...

Publication Years

Zbl 64.35501
Grothendieck, Alexandre
Produits tensoriels topologiques et espaces nucléaires. (Topological tensor products and nuclear spaces.) (French)
Mem. Am. Math. Soc. 16, 190 p., 140 p. (1955).

Es kann auf die ausführliche Besprechung der Vorankündigung in Ann. Inst. Fourier 4, 73–112 (1954; Zbl 55.09705) hingewiesen werden. *Gottfried Köthe*

Zbl 55.09705
Grothendieck, Alexander
Résumé des résultats essentiels dans la théorie des produits tensoriels topologiques et des espaces nucléaires. (Summary and essential results in the theory of topological tensor products and nuclear spaces.) (French)
Ann. Inst. Fourier 4, 73-112 (1952).

Die vorliegende Arbeit enthält eine Zusammenstellung der Hauptresultate der in den Memoirs of the Amer. Math. Soc. erscheinenden Thèse des Verf. ohne Beweise. Kapitel I enthält die Theorie der topologischen Tensorprodukte lokalkonvexer Räume. Sind E, F zwei lokalkonvexe Räume, so erhält man auf dem gewöhnlichen Tensorprodukt $E \otimes F$ eine lokalkonvexe Topologie, wenn man als Nullumgebungssystem die Mengen $\Gamma(U \otimes V)$ nimmt, die

Author profile:

Grothendieck, Alexander

Spellings: Grothendieck, A. [74] Grothendieck, Alexander [26] Grothendieck, Alexandre [16] Grothendieck, A. [1]
Author-Id: grothendieck.alexander
Publications: 117 including 32 Book(s) and 41 Journal Article(s)

MSC 2010

55	14 · Algebraic geometry
15	00 · General mathematics
13	18 · Category theory, homological algebra
9	46 · Functional analysis
4	01 · History; biography

more ...

Journals

6	Publications Mathématiques
4	Comptes Rendus de l'Académie des Sciences. Paris
3	Bulletin de la Société Mathématique de France
3	Canadian Journal of Mathematics
2	American Journal of Mathematics

more ...

Co-Authors

14	Demazure, Michel
5	Dieudonné, Jean Alexandre
3	Verdier, Jean-Louis
1	Berthelot, Pierre
1	Illusie, Luc

more ...

Publication Years

absolutkonvexen Hüllen der Tensorprodukte zweier Nullumgebungen U und V aus E bzw. F. Als projektives Tensorprodukt $E\widehat{\otimes}F$ wird die vollständige Hülle von $E\otimes F$ bezeichnet. Im Fall zweier (F)-Räume hat jedes Element von $E\widehat{\otimes}F$ die Form $\sum_{i=1}^{\infty}\lambda_i x_i \otimes y_i$, wobei x_i bzw. y_i beschränkte Folgen aus E bzw. F sind und $\sum|\lambda_i| < \infty$. Es wird $\mathfrak{L}^1(\mu)\widehat{\otimes}E$ für einen (B)-Raum E gleich dem Raum $\mathfrak{L}_E^1(\mu)$ der μ-integrablen Abbildungen eines lokalkompakten Raumes M in den (B)-Raum E. Sind E, F zwei (B)-Räume, so ist $E \otimes F$ linearer Teilraum des (B)-Raumes $B(E', F')$ der auf $E' \times F'$ stetigen Bilinearformen. Die abgeschlossene Hülle von $E \otimes F$ in $B(E', F')$ in der induzierten Norm wird als $E\widehat{\otimes}F$ bezeichnet. $E\widehat{\otimes}F$ läßt sich auch für beliebige lokalkonvexe E, F erklären.

Die Frage, ob die stetige lineare Abbildung $E\widehat{\otimes}F \to E\widehat{\otimes}F$ eineindeutig ist, ist ungeklärt. Sind E, F (B)-Räume, so können die Elemente von $E'\widehat{\otimes}F$ als stetige lineare Abbildungen von E in F aufgefaßt werden. Eine solche Abbildung heißt nuklear. Für beliebige lokalkonvexe E, F wird eine Abbildung von E in F als nuklear bezeichnet, wenn sie als Produkt $\gamma \, \beta \, \alpha$ geschrieben werden kann, α stetige Abbildung von E in einen (B)-Raum E_1, β nukleare Abbildung von E_1 in einen (B)-Raum E_2, γ stetige Abbildung von E_2 in F. Eine nukleare Abbildung ist stets kompakt. Eine die nuklearen Abbildungen umfassende

Klasse bilden die integralen Abbildungen von E in F, das sind solche, die sich als stetige

Linearfunktionen auf $\widehat{E \otimes F}'$ auffassen lassen. Sie gestatten Darstellungen als Integraloperationen. Das Produkt zweier integralen Abbildungen ist eine nukleare Abbildung. Eine dritte Sorte Abbildungen, die studiert wird, sind die Fredholmschen Abbildungen, die ebenfalls die nuklearen umfassen und in vielen Fällen mit ihnen zusammenfallen. Für das bekannte Problem, ob jede kompakte Abbildung Limes von Abbildungen endlichen Ranges ist, werden zahlreiche äquivalente Formulierungen gegeben, es wird in vielen Fällen auch gelöst. Im Kapitel II werden die nuklearen Räume studiert, das

sind lokalkonvexe Räume E, für die $\widehat{E \otimes F} = \widehat{E \otimes F}$ für jedes lokalkonvexe F gilt. Jede lineare stetige Abbildung eines nuklearen Raumes E in einen (B)-Raum ist nuklear. Ist E nuklear und quasivollständig, so ist E ein (M)-Raum, speziell also reflexiv. Jeder Teilraum und jeder Quotientenraum ist wieder nuklear, ebenso das topologische Produkt beliebig vieler und die topologische Summe abzählbar vieler nuklearer Räume. Die bekannten Räume $\mathfrak{E}$, $\mathfrak{E}'$, $\mathfrak{D}$, $\mathfrak{D}'$, S, S', O_M, O_0' der Schwartzschen Distributionentheorie sind nuklear, ebenso der Raum H der analytischen Funktionen auf einer komplexen Mannigfaltigkeit. Die Eigenschaften von $\widehat{E \otimes F}$ im Fall eines nuklearen E und eines (F)-Raumes F werden untersucht. Stetige Abbildungen von E in F der Form $\sum \lambda_i x_i' \otimes y_i$ mit $(\lambda_i) \in l^p$, $0 < p \leq 1$, x_i bzw. y_i je aus einer kompakten Teilmenge von E' bzw. F heißen von p-ter Potenz summierbare Fredholmsche Abbildungen. Eine Fredholmsche Abbildung hat eine Fredholmsche Determinante, die eine ganze Funktion ist. Sätze über Zusammenhänge zwischen der Ordnung der Fredholmschen Determinante und der Potenz, in der die zugehörige Fredholmsche Abbildung summierbar ist.   *Gottfried Köthe*

Zbl 70.16302
Hirzebruch, Friedrich
Neue topologische Methoden in der algebraischen Geometrie.
Ergebnisse der Mathematik und ihrer Grenzgebiete. 9.
Berlin-Göttingen-Heidelberg: Springer-Verlag. VIII, 165 S. (1956).

Author profile:

Hirzebruch, Friedrich Ernst Peter

Spellings: Hirzebruch, Friedrich [112] Hirzebruch, F. [10] Hirzebruch, Friedrich Ernst Peter [1] Hirzebruch, Friedrich E.P. [1] Hirzebruch, Friedrich E. P. [1]
Author-Id: hirzebruch.friedrich-ernst-peter
Publications: 125 including 31 Book(s) and 57 Journal Article(s)

MSC 2010

41	14 · Algebraic geometry
26	11 · Number theory
24	01 · History; biography
23	57 · Manifolds and cell complexes
14	00 · General mathematics

more ...

Journals

7	Mathematische Annalen
4	Notices of the American Mathematical Society
3	L'Enseignement Mathématique. IIe Série
2	American Journal of Mathematics
2	Astérisque.

more ...

Co-Authors

10	Atiyah, Michael Francis
8	Remmert, Reinhold
5	Ebbinghaus, Heinz-Dieter
5	Köcher, Max
5	Mainzer, Klaus

more ...

Publication Years

Il fallait une certaine audace pour publier sous forme de livre les brillants résultats que l'A. a obtenus si récemment. On peut cependant affirmer que la tentative est pleinement réussie. L'ouvrage présenté donne un exposé – unique dans la littérature existante – de la théorie des faisceaux et des espaces fibrés, vue du point de vue le plus moderne, et son intérêt n'est pas moindre pour le Topologue que pour le Géométre Algébriste. Le chapitre I – presque la moitié de l'ouvrage – expose les théories de base: Calcul formel (dû à l'A.) qui associe à toute série entière une suite de polynomes. Théorie des faisceaux (cohomologie d'un espace à valeur dans un faisceau, suite exacte, etc.); théorie des espaces fibrés à fibre vectorielle. Ici la théorie de l'isomorphisme des fibrés est faite, du point du vue moderne, en associant au fibré une classe du groupe H de la base X à valeurs dans le faisceau des germes d'applications de X dans le groupe de structure G. Puis intervient le procédé, fondamental pour la suite, de réduction du groupe de structure G à un sous-groupe fermé: diagonalisation du groupe des rotations $SO(k)$ par passage au fibré induit sur le fibré sur X en variétés de drapeaux $F(k) = SO(k)/\Delta^k$. Ce procédé permet de définir les classes caractéristiques de Chern et Pontrjagin par la théorie de Borel-Serre du tore maximal, ce qui conduit à des définitions beaucoup plus maniables que les définitions classiques: les théorèmes de "dualité", notamment, se démontrent immédiatement, et toute équivoque sur le signe des classes de Chern est écartée. Le chapitre II donne ce qu'il est necessaire de savoir Sur les nombres caractéristiques de Pontrjagin et la théorie du "cobordisme" afin d'établir le théorème de l'index, qui permet d'exprimer l'index $\tau(V)$ d'une variété orientée en fonction de ses nombres de Pontrjagin. Définition et propriétés des indices virtuels, associés à un système de classes de cohomologie de $H^2(V; Z)$. Le chapitre III donne, dans une variété presque-complexe, la définition du genre de Todd, et du genre de Todd virtuel, absolu ou relatif à un fibré. Propriétés multiplicatives et arithmétiques du genre de Todd, identité fonctionnelles auxquelles satisfait le genre T_y. Invariance du genre de Todd par passage au fibré en variétés de drapeaux, avec, pour terminer, l'expression du genre de Todd d'une variete à fibré diagonal en fonction d'indices virtuels. Le Chapitre IV en arrive au point essentiel: Définition de la caractéristique χ d'une variété complexe compacte à valeurs dans un faisceau comme somme alternée des rangs (finis) des groupes de cohomologie $H^i(V; F)$; T^p désignant le faisceau des p-formes holomorphes, on pose $\chi^p = \chi(V; F \circ T^p)$ et $\chi_y = \sum y^p \chi^p$. On a alors $\chi_{-1} = $ caractéristique d'Euler-Poincaré, et, pour une variété kählérienne, $\chi_1(V) = \tau(V)$ (Th. dû à Hodge) pour le faisceau des fonctions holomorphes. La formule des 4 termes de Kodaira-Spencer (qui est l'équivalent, pour les faisceaux, de la dualité de Whitney), permet de définir le comportement de la caractéristique χ_y par restriction à un diviseur sans singularités. Par extension de cette formule à un diviseur arbitraire, on définit des caractéristiques χ_y virtuelles, qui ont les mêmes propriétés formelles que le genre de Todd T_y. Puls l'A. introduit les théorèmes fondamentaux de Kodaira sur la caractérisation des diviseurs positifs et des variétés algébriques. Le fait que taut diviseur est différence de deux diviseurs positifs et sans singularités permet de passer du cas "réel" au cas virtuel: on montre ainsi que la caractéristique χ_1 est égale à l'index τ même dans le cas virtuel, et les formules établies pour T_y et χ_y dans une variété diagonalisable (au sens analytique) permettent d'identifier le genre de Todd $T_0(V)$ et le genre arithmétique $\chi_0(V)$. On passe de là au cas général d'une variété algébrique V en substituant à V le fibré V^Δ dont la fibre est la variété de drapeaux (V^Δ est encore algébrique, et $\chi_0(V^\Delta) = \chi_0(V)$). La démonstration proposée souffre peut-être d'un abus de formalisme, et pourra peut-être être simplifiée ultérieurement, au moins en présentation (On a l'impression, par exemple, que dans la théorie de la diagonalisation, le fibré des vecteurs tangents aux fibres variétés de drapeaux ne joue qu'un rôle parasite, qui devrait pouvoir être éliminé). Finalement, l'A. aboutit à une formule générale qui donne le χ d'une variété algébrique V à valeurs dans le faisceau des germes de sections holomorphes d un fibré analytique G comme un polynome par rapport aux classes de Chern de V et du fibré G. Un paragraphe final fait le raccord avec la théorie classique de Riemann-Roch. Le lecteur peut ainsi se rendre compte du progrès considérable apporté dans cette branche de la Géométrie Algébrique par les méthodes extrêmement générales de l'A. *René Thom*

Zbl 99.03903
Erdős, Pál; Rényi, Alfréd
A probabilistic approach to problems of diophantine approximation.
Ill. J. Math. 1, 303-315 (1957).

Author profile:

Erdős, Paul

Spellings: Erdős, Paul [1009] Erdős, Pál [444] Erdős, P. [79] Erdős, Paul [7] Erdős, P. [4] Erdős, Pál [2] Erdős, P. [2] Erdős, Paul [1]
Author-Id: erdos.paul
Publications: 1548 including 11 Book(s) and 1250 Journal Article(s)

MSC 2010

630	**11** · Number theory
597	**05** · Combinatorics
172	**00** · General mathematics
88	**03** · Mathematical logic
76	**52** · Convex and discrete geometry

more ...

Journals

80	Discrete Mathematics
67	Journal of the London Mathematical Society
51	Acta Arithmetica
38	Acta Mathematica Academiae Scientiarum Hungaricae
38	Bulletin of the American Mathematical Society

more ...

Co-Authors

61	Sárközy, András
56	Hajnal, András
49	Faudree, Ralph J.
40	Schelp, Richard H.
39	Turán, Pál

more ...

Publication Years

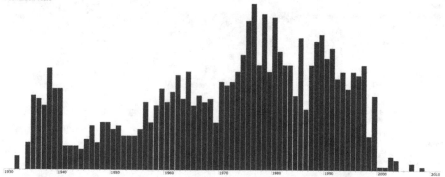

| 1930 | 1940 | 1950 | 1960 | 1970 | 1980 | 1990 | 2000 | 2010 |

Es seien stets $z_j = e^{i\varphi_j}$ ($0 \le \varphi_j < 2\pi; j = 1, \ldots, n$) komplexe Zahlen vom Betrag 1, $S_k = \sum_{j=1}^{n} z_j^k$ (k natürliche Zahl). Wir bezeichnen eine Menge $(z_1, \ldots, z_n)$ stets mit Z_n. Dann zeigen die Verff. folgende Sätze.

Satz 1: Es gibt für jedes $n \ge 2$ ein Z_n, so daß $|S_k| < \sqrt{6n \log(k+1)}$ ($k = 1, 2, \ldots$). Ein ähnliches Resultat gilt, wenn k reell beliebig ist. Satz 2: Es gibt für $n \ge 2$ stets ein Z_n, so daß $|S_k| < cn$ für alle k mit $1 \le k < \frac{1}{4} \exp(nc^2/2)$ für jedes c in $0 < c < 1$.

Dagegen (Satz 6) ist für alle Z_n und $2 \le c \le n - 1$,

$$\max_{1\le k\le 2n(4\pi n\,\sqrt{c+2})^{c+2}}|S_k|\ge c.$$

Satz 3: Für $n \ge 10$ und $0 < \varepsilon < 1/16$ gibt es Z_n, so daß

$$|S_k| < n(1-\varepsilon), \qquad (1 \le k \le (16n\varepsilon^{n-1})^{-1/2}).$$

Der Beweis dieser Sätze stützt sich auf das Lemma: Sind $\varphi_1,\dots,\varphi_n$ unabhängige zufällige Variable, gleich verteilt auf $(0,2\pi)$, dann ist für jedes c mit $0 < c < 1$, $P(|S_k| \ge cn) \le 4e^{-c^2 n/2}$ (P: Wahrscheinlichkeit). In den Sätzen kann noch angenommen werden, daß die z_j Einheitswurzeln vom Grad p (p Primzahl $> p_0(n)$) sind. In Verbindung mit dem Satz von *Erdős-Turán* aus der Theorie der Gleichverteilung (Zbl 31.25402) folgt aus Satz 1: Es gibt ein Z_n, so daß für die Anzahl $N_n^{(k)}(\alpha,\beta)$ der $z_1^k,\dots,z_n^k$ auf dem Bogen (α,β) des Einheitskreises gilt

$$|N_n^{(k)}(\alpha,\beta)/n - (\beta-\alpha)/2\pi| < c\,\sqrt{\delta}[\log(e\delta)]^{3/2} \text{ für } k \le e^{\delta n} - 2$$

$0 < \delta < 1$ (c absolute Konstante). Zur Bedeutung dieser Sätze siehe das Buch von *P.Turán* (Zbl 52.04601). *E. Hlawka*

Author profile:

Rényi, Alfréd

Spellings: Rényi, Alfréd [189] Rényi, A. [15] Renyi, Alfred [12] Renyi, A. [5] Renyi, Alfréd [2] Reni, A.A. [2] Rényi, Alfred [1] Rény, Alfréd [1] Rény, A. [1]
Author-Id: renyi.alfred
Publications: 228 including 24 Book(s) and 177 Journal Article(s)

MSC 2010

28	60 · Probability theory and stochastic processes
18	05 · Combinatorics
16	11 · Number theory
14	00 · General mathematics
9	01 · History; biography

more ...

Journals

28	Acta Mathematica Academiae Scientiarum Hungaricae
10	Publicationes Mathematicae
6	Annales Universitatis Scientiarum Budapestinensis de Rolando Eötvös Nominatae. Sectio Mathematica
6	Matematikai Lapok
6	Selected Translations in Mathematical Statistics and Probability

more ...

Co-Authors

34	Erdős, Paul
4	Rényi, Catherine
3	Prékopa, András
3	Sulanke, Rolf
2	Aczél, János

more ...

Publication Years

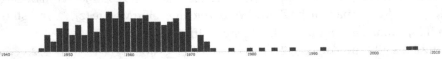

Zbl 80.03502
Hlawka, Edmund
Zur Theorie der diophantischen Approximationen.
Österr. Akad. Wiss., Math.-Naturw. Kl., Anz. 1958, 25-49 (1958).

In this note the author generalizes measure-theoretic results of P. Erdős and A. Renyi] about the behaviour of sums of powers of complex numbers [Ill. J. Math. 1, 303–315 (1957; Zbl 99.03903)]. The proofs are fairly straightforward and are on somewhat different lines from those of Erdős and Renyi. Of the many theorems given, the following may be taken as a representative sample: Let $f_1(x), \ldots, f_N(x)$ be functions defined on a compact space X with a denumerable basis and taking values in the field of complex numbers. Let a_k ($1 < k < \infty$) be a sequence of complex numbers and suppose that $\sum |a_k|^2 = \alpha < \infty$, $\sum |a_k| < \infty$. Let $\delta > 0$ be such that $N < 3^{-10} \exp(\delta^2/\alpha)$. Then there is a sequence $x_k \in X$ such that $\sum |a_k f_j(x_k)| < 3\delta K$ $1 \leq j \leq N$, where $K = \text{Max}_{1 \leq j \leq N} \sup_{x \in X} |f_j(x)|$.   J. W. S Cassels

Author profile:

Hlawka, Edmund

Spellings: Hlawka, Edmund [122] Hlawka, E. [50] Hlawka, M.Edmund [1]
Author-Id: hlawka.edmund
Publications: 173 including 14 Book(s) and 139 Journal Article(s)

MSC 2010

95	11 · Number theory
22	01 · History; biography
13	65 · Numerical analysis
6	30 · Functions of a complex variable
6	42 · Fourier analysis

more ...

Journals

20	Sitzungsberichte. Abteilung II. Österreichische Akademie der Wissenschaften, Mathematisch-Naturwissenschaftliche Klasse
18	Monatshefte für Mathematik
9	Österreichische Akademie der Wissenschaften, Mathematisch-Naturwissenschaftliche Klasse. Sitzungsberichte. Abteilung II.
8	Acta Arithmetica
6	Monatshefte für Mathematik und Physik

more ...

Co-Authors

4	Binder, Christa
4	Schoißengeier, Johannes
2	Firneis, Friedrich J.
2	Muck, Rudolf
2	Taschner, Rudolf J.

more ...

Publication Years

76

Zbl 128.15302
Freudenthal, Hans
Beziehungen der $\mathfrak{E}_7$ und $\mathfrak{E}_8$ zur Oktavenebene. V-IX.
Nederl. Akad. Wet., Proc., Ser. A 62, 165-179, 180-191, 192-201, 447-465, 466-474 (1959).

Die vorliegende umfangreiche Arbeit verfolgt wohl das Hauptziel, die Lieschen einfachen Ausnahmegruppen $\mathfrak{F}_4$, $\mathfrak{E}_6$, $\mathfrak{E}_7$ und $\mathfrak{E}_8$ als Symmetriegruppen einer über $\mathbb{R}$, $\mathbb{C}$, $\mathbb{H}$, $\mathbb{O}$ (reellen, komplexen, Quaternionen, Oktavzahlen) gebildeten besonderen, metasymplektisch genannten Geometrie (MSG) zu erklären. Hierzu vergleiche man auch *J. Tits* [Acad. Roy. Belgique, Cl. Sci., Mém., Coll. 8 No. 3 (1955; Zbl 67.12301); Bull. Soc. Math. Belg. 8, 48–81 (1956; Zbl 72.38202)]. Die MSG enthält 4 Raumelemente: Punkt, Gerade, Ebene, 5 dimensionales sogenanntes Symplekton mit 3 binären Punktrelationen: Verbundenheit $\Rightarrow$ Verflechtung (beide Punkte gehören zu passendem Symplekton) $\Rightarrow$ Scharnierung (beide Punkte sind mit passendem Punkt verbunden). Ähnlich heißen zwei Symplekta verbunden (wenn ihr Durchschnitt eine Ebene ist), $\Rightarrow$ verflochten (wenn beide einen passenden Punkt enthalten), $\Rightarrow$ scharnierend (wenn beide mit passendem Symplekton verbunden sind). Die MSG ist hier, vermutlich provisorisch, zunächst mehr rechnerisch im Sinne der analytischen Geometrie definiert. Viele interessante, in geometrischer Sprache formulierte Beziehungen werden infolgedessen oft durch mühsame Rechnungen hergeleitet. Wegen der noch provisorischen Fassung der sicher schönen und wichtigen MSG soll in diesem Referat auf Wiedergabe von Einzelheiten verzichtet werden. Eine Liste von Berichtigungen zu den vorangegangenen Teilen I–IV [Indag. Math. 16, 218–230; 363–368 (1954; Zbl 55.02001; Zbl 58.26101); Nederl. Akad. Wet., Proc., Ser. A 58, 151–157, 277–285 (1955; Zbl 68.14104)] ist auf S. 167–168 angegeben. [Teil X und XI vgl. Nederl. Akad. Wet., Proc., Ser. A 66, 457–471, 472–487 (1963; Zbl 123.13501)]. *Ernst Witt*

Author profile:

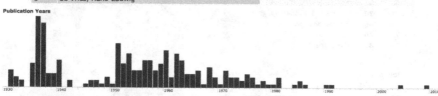

Zbl 91.05802
Rudin, Walter
Trigonometric series with gaps.
J. Math. Mech. 9, 203-227 (1960).

Es sei E eine nicht-leere Menge ganzer Zahlen. Eine Funktion $f \in L_1(-\pi, \pi)$ (trigonometrisches Polynom) heiße E-Funktion [E-Polynom], wenn die Fourierkoeffizienten

$$\hat{f}(n) = \frac{1}{2\pi} \int_{-\pi}^{\pi} f(t) \exp(int)dt$$

für $n \notin E$ null sind. Die Menge E heiße Sidonsch, wenn es eine Konstante B gibt, für welche

$$\sum_{-\infty}^{\infty} |\hat{f}(n)| \le B\|f\|_\infty$$

für alle E-Polynome f. Wir bezeichnen den Raum aller E-Funktionen in $L_p(-\pi, \pi)$ ($0 < p < \infty$) als $L_{p,E}$. Für $0 < s < \infty$ heißt eine Menge E vom Typus

$\Lambda(s)$, wenn es ein r $(0 < r < s)$, gibt, so daß $L_{r,E} = L_{s,E}$. Eine Reihe von Sätzen über E-Funktionen werden gegeben. Folgende sind typisch.

(A) Die Menge E ist Sidonsch, wenn es eine Konstante $\delta > 0$ mit folgender Eigenschaft gibt: Für jede Funktion b auf E, $b(n) = 1$, gibt es ein Maß μ auf $[-\pi, \pi]$ mit $\hat{\mu}(n) - b(n) < 1 - \delta$ für alle $n \in E$.

(B) Jede Sidonsche Menge ist vomTypus $\Lambda(s)$ für alle $s > 0$. In der Tat, ist f ein E-Polynom, dann gilt $\|f\|_s \leq B\sqrt{s}\|f\|_2$, $2 < s < \infty$, $\|f\|_2 \leq B\|f\|_1$, wobei B die Konstante in der Deinition einer Sidonschen Menge ist.

Verschiedene Eigenschaften von $\Lambda(s)$-Mengen werden untersucht und einige frühere Resultate werden wiedergeben. Siehe z.B. *H. S. Zuckerman* und der Referent [Trans. Am. Math. Soc. 93, 1–19 (1959; Zbl 141.12601)]. *Edwin Hewitt*

Author profile:

Rudin, Walter

Spellings: Rudin, Walter [149] Rudin, W. [39]
Author-Id: rudin.walter
Publications: 188 including 31 Book(s) and 147 Journal Article(s)

MSC 2010

50	**32** · Functions of several complex variables and analytic spaces
31	**30** · Functions of a complex variable
24	**46** · Functional analysis
22	**26** · Real functions
11	**31** · Potential theory

more ...

Journals

24	Proceedings of the American Mathematical Society
13	Bulletin of the American Mathematical Society
13	Duke Mathematical Journal
12	American Mathematical Monthly
8	Transactions of the American Mathematical Society

more ...

Co-Authors

10	Ahern, Patrick B.
5	Nagel, Alexander
5	Rosay, Jean-Pierre
2	Kahane, Jean-Pierre
2	Katznelson, Yitzhak

more ...

Publication Years

| 1940 | 1950 | 1960 | 1970 | 1980 | 1990 | 2000 | 2010 |

Zbl 102.03601
Lewis, Donald J.; Mahler, K.
On the representation of integers by binary forms.
Acta Arith. 6, 333-363 (1961).

Author profile:

Lewis, Donald-J.

Spellings: Lewis, D.J. [58] Lewis, Donald J. [2] Lewis, Donald-J. [1]
Author-Id: lewis.donald-j
Publications: 61 including 4 Book(s) and 47 Journal Article(s)

MSC 2010

25	11 · Number theory
8	14 · Algebraic geometry
3	12 · Field theory and polynomials
2	00 · General mathematics
2	01 · History; biography

more ...

Journals

7	Acta Arithmetica
5	Michigan Mathematical Journal
5	The Quarterly Journal of Mathematics. Oxford Second Series
3	American Journal of Mathematics
3	Mathematika

more ...

Co-Authors

16	Davenport, Harold
5	Birch, Bryan J.
5	Chowla, Sarvadaman
5	Schinzel, Andrzej
4	Dunton, M.

more ...

Publication Years

The most striking result is the following consequence of Theorem 2. Let $F(x, y)$ be a binary form of degree n with rational integral coefficients and nonzero discriminant such that $F(1, 0) \neq 0$, $F(0, 1) \neq 0$. Let a be the height of F (= maximum of the absolute value of the coefficients and let $p_1, \ldots, p_t$ be a fixed set of t primes. Then

(i) there are at most $c_1 (an)^{c_2 n^{1/2}} + c_3 n^{t+1}$ coprime pairs of integers x, y for which $F(x, y)$ has no prime factors distinct from $p_1, \ldots, p_t$;

(ii) there are at most $(c_3 n)^{t+1}$ coprime pairs of integers x, y such that $F(x, y)$ is greater than a certain constant depending an a and n and for which $F(x, y)$ has no prime factors distinct from $p_1, \ldots, p_t$.

(iii) Let p be a prime. Then if p is sufficiently large there are at most $c_4 n^2$ coprime pairs of integers for which $\pm F(x, y)$ is equal to p or a power of p. Here c_1, c_2, c_3, c_4 are constants which can be given explicitly and are not too large.

80

Theorem 2 itself gives better but very elaborate estimates.

Statement (i) above gives, in particular, the upper bound $c_1(an)^{c_2 n^{1/2}} + c_3 n^{t+1}$ for the number of solutions of $F(x, y) = m$, where t is the number of prime factors of m, and this is much better than the estimate obtained by H. Davenporth and K. F. Roth [Mathematika 2, 160–167 (1955; Zbl 66.29302)].

Author profile:

Mahler, Kurt

Spellings: Mahler, K. [179] Mahler, Kurt [104]
Author-Id: mahler.kurt
Publications: 283 including 5 Book(s) and 273 Journal Article(s)

MSC 2010

58	11 · Number theory
12	30 · Functions of a complex variable
8	12 · Field theory and polynomials
5	39 · Difference and functional equations
4	01 · History; biography

more ...

Journals

30	Journal of the London Mathematical Society
18	Proceedings. Akadamie van Wetenschappen Amsterdam
13	Bulletin of the Australian Mathematical Society
13	Mathematische Annalen
12	Acta Mathematica

more ...

Co-Authors

4	Erdős, Paul
3	Popken, Jan
2	Billing, Gert Due
2	Ledermann, Walter
2	Segre, Beniamino

Publication Years

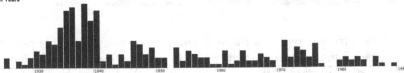

The proof is an elaboration of an earlier one of K. Mahler [Math. Ann. 107, 691–730 (1933; Zbl 6.10502; JFM 59.0220.01)] of a less precise result (which as the authors point out is, however in general much stronger than that of Davenport and Roth), and is related closely to Mahler's work an the p-adic Thue–Siegel theorem. Of independent interest are Lemma 1 giving a lower bound for $f(z)$, where f is a polynomial wich complex coefficients, in terms of the height, discriminant and degree of f and of the minimum distance of z from a zero of f which is stronger than that given by N. I. Fel'dman [Izv. Akad. Nauk SSSR, Ser. Mat. 15, 53–74 (1951; Zbl 42.04802)] and the p-adic

analogue (Lemma 2) which improves an estimate of *F. Kasch* and *B. Volkmann* [Math. Z. 72, 367–378 (1960; Zbl 91.04801)].
Using these the authors obtain a lower bound for

$$|F(x,y)| \prod_{\tau=1}^{t} |F(x,y)|_{p_\tau}$$

in terms of the minimum distance of x/y from the complex and p-adic zeros of $F(x,1) = 0$. When the only prime factors of $F(x,y)$ are $p_1, \ldots, p_t$ then the expression (*) is unity. Hence the machinery of the p-adic Thue–Siegel theorem applies. There is an elaborate and ingenious subdivision into cases. There is also a rather elaborate generalization of Theorem 2 (Theorem 3) in which, roughly speaking, although $F(x,y)$ is not entirely composed of $p_1, \ldots, p_t$, the product of the primes dividing $F(x,y)$ other than these is of lower order of magnitude.

J. W. S. Cassels

Zbl 118.36206
Grothendieck, A.
Éléments de géométrie algébrique. I: Le langage des schemas. II: Étude globale élémentaire de quelques classe de morphismes. III: Étude cohomologique des faisceaux cohérents (premiere partie)
Publ. Math., Inst. Hautes Étud. Sci. 4, 1-228 (1960); ibid. 8, 1-222 (1961); ibid. 11, 349-511 (1962).

Von dieser großangelegten Neubegründung der algebraischen Geometrie in den "Elementen" (abgekürzt EGA) – in der Einleitung werden vorläufig die Titel von 13 Kapiteln angegeben – sind bisher dreieinhalb Kapitel mit einem Umfang von annähernd 1000 Seiten erschienen. Besser als ein Referat können über Zielsetzung, Ergebnisse und Methoden dieser umfang- und inhaltsreichen Darstellung die vom Verf. in den Jahren 1957–62 im Séminaire Bourbaki gehaltenen 8 Vorträge Auskunft geben, die inzwischen unter dem Titel "Fondements de la géométrie algébrique, Extraits du Séminaire Bourbaki 1957–62, Paris 1962" gesammelt erschienen sind, sowie die Vorträge vom Verf. [Proc. internat. Congr. Math. 14–21 Aug. 1958, 103–118 (1960)] und von *J. P. Serre* in [Proc. internat. Congr. Math. 15–22 Aug. 1962, 190–196 (1963)]. In einem Kapitel 0, jeweils den Kapiteln 1, 3,1. und 4,1 vorangestellt, werden Ergebnisse der kommutativen und der homologischen Algebra sowie der Garbentheorie als Hilfsmittel zusammengetragen. Man vergleiche auch die Errata und Addenda am Ende von Kap. 2, Kap. 3,2 und Kap. 4,1.
Grob gesprochen treten in der Grothendieckschen Theorie der Schemata an die Stelle der Polynomalgebren und ihrer Restklassenringe in der klassischen algebraischen Geometrie beliebige kommutative Ringe, zuweilen mit "Endlichkeitsvoraussetzungen", z.B. Noethersche Ringe. Jedenfalls sind diese Voraussetzungen erfüllt für die in der Zahlentheorie wichtigen Ringe. Es ist in diesem Rahmen auch möglich, Schemata über lokalen Ringen, sowie solche mit nilpotenten Elementen in den Ringen der Strukturgarbe zu be-

trachten, was technisch von höchster Bedeutung ist (vgl. den Übergang von gewöhnlichen zu formellen Schemata). Der Zusammenhang zwischen dem Lokalen und dem Globalen formuliert sich wie in der analytischen Geometrie am bequemsten in der Sprache der Garbentheorie. Eingeführt in die abstrakte algebraische Geometrie wurde die letztere zusammen mit den Methoden der algebraischen Topologie ausgehend von der Zariski Topologie einer Varietät von *J.-P. Serre* in der für die moderne algebraische Geometrie richtungsweisenden Arbeit FAC [Ann. Math. (2) 61, 197–278 (1955; Zbl 67.16201)]. Ferner macht die Darstellung ausgiebig Gebrauch von der "funktoriellen Sprache", wie sie bei *H. Cartan* und *S. Eilenberg*, "Homological algebra" (Princeton 1956) und *A. Grothendieck* [Sur quelques points d'algebre homologique. Traduction du français par B. B. Venkov, A. V. Rukolaine und B. V. Stepanov, Moskau: Verlag für ausländische Literatur (1961; Zbl 118.26103)] u.a. vor allem in der algebraischen Topologie entwickelt wurde. Dies macht die Lektüre der EGA für einen darin ungeübten Leser bisweilen etwas mühsam. Er wird aber schließlich für seine Mühen durch die Fülle der Resultate und vor allem der gewonnenen neuen Gesichtspunkte reich belohnt werden. Im einzelnen enthält Kapitel 0 aus der kommutativen Algebra Grundlegendes über Quotientenringe und Quotientenmoduln, irreduzible und Noethersche topologische Räume, flache und treu flache Moduln und Ringerweiterungen und im Hinblick auf die formellen Schemata über zulässige (anneaux admissible) und adische Ringe (anneaux adique), was man inzwischen zum Teil auch bei *N. Bourbaki*, Éléments de mathématique, Algèbre commutative, chap. 1–4 (vgl. u.a. dies Zbl 108.04002) nachlesen kann. Dabei heißt ein Ring A zulässig, wenn er vollständiger topologischer Ring ist mit einem Fundamentalsystem von Nullumgebungen bestehend aus Idealen und einem offenen Ideal J, das topologisch nilpotent ist. Ein solches Ideal heißt ein Definitionsideal. Die Definitionsideale bilden eine Basis für die Nullumgebungen. Man kann daher A auch erhalten als projektiven Limes $A = \varprojlim A_\lambda$ eines projektiven Systems $(A_\lambda, u_{\lambda\mu})$ von diskreten Ringen A_λ mit einer nach rechts gefilterten Indexmenge L, die ein kleinstes Element 0 besitzt, so daß

i) $u_\lambda : A \to A_\lambda$ surjektiv und

ii) der Kern von $u_{0\lambda} : A_\lambda \to A_0$ nilpotent ist.

Spezialfälle sind die lokalen linear kompakten Ringe. Ein zulässiger Ring A heißt J-adisch wenn J ein Definitionsideal ist, dessen Potenzen J^n ein Fundamentalsystem von Nullumgebungen bilden. [Proposition (7.2.4.) ist in der angegebenen Form falsch. Siehe (Err$_{III}$, 3) in Kapitel 3,2 p. 87]. Ferner enthält Kapitel 0 Grundlegendes aus der Garbentheorie und der Theorie der geringten Räume. Erwähnenswert ist vielleicht die "richtige" Definition des "reziproken Bildes" als links adjungierter Funktor des Funktors "direktes Bild".

Kapitel 1 entwickelt die Sprache der Schemata beginnend mit den affinen Schemata. Für einen kommutativen Ring A ist Spec(A) die Menge der

Primideale von A versehen mit der Zariski Topologie. Das affine Schema Spec(A) ist nun der eben definierte topologische Raum, mit einer durch A definierten Garbe von Ringen $\tilde{A}$, also Strukturgarbe. Die Faser über dem Punkt $\mathfrak{p} \in$ Spec(A) ist gleich dem Quotientenring $A_{\mathfrak{p}}$ also lokal. Ein Präschema $\tilde{X}$ ist ein geringter topologischer Raum, der eine Überdeckung durch offene affine Mengen V_i besitzt. Dabei heißt $V \subset X$ offen affin, wenn die offene Menge V mit der von X induzierten geringten Struktur ein affines Schema im obigen Sinne ist. Ein Morphismus zwischen Präschemata ist ein Morphismus zwischen den geringten Räumen, der in den Fasern lokale Homomorphismen induziert. Mit dieser Definition ist die Kategorie der affinen Schemata dual zur Kategorie der kommutativen Ringe. Die Kategorie der quasikohärenten Garben auf Spec(A) ist äquivalent zur Kategorie der A-Moduln, die der kohärenten zur Kategorie der endlich erzeugten A-Moduln, falls A Noethersch. In der Kategorie der Präschemata existieren endliche Faserprodukte. Deren Definition ist typisch für alle allgemeinen Konstruktionen in Kategorien: Man überträgt in der Kategorie der Mengen definierte Begriffe mit Hilfe von $\underset{K}{\text{Hom}}(\, , \,)$ in die beliebige Kategorie K (hier die der Präschemata): Für Morphismen $X \to S$, $Y \to S$ ist das Faserprodukt über S $X \underset{S}{\times} Y$ dasjenige Präschema $\in K$, für welches für alle $T \in K$, $\underset{K}{\text{Hom}}(T, X \underset{S}{\times} Y) \cong \text{Hom}(T, X) \underset{\text{Hom }(T,S)}{\times} \text{Hom}(T, Y)$ funktoriell in T ist, wo rechts das Faserprodukt in der Kategorie der Mengen steht. Falls es ein solches Präschema gibt, ist es durch diese Forderung bis auf kanonische Isomorphie eindeutig bestimmt. Man sagt auch $X \underset{S}{\times} Y$ repräsentiert den Funktor $T \rightsquigarrow \text{Hom}(T, X) \underset{\text{Hom }(}{\times} T, S) \text{Hom}(T, Y)$ von der gegebenen Kategorie der Präschemata K in die Kategorie der Mengen E. Das Faserprodukt liefert auch die Definition der Fasern eines Morphismus $f : X \to Y$ über Punkten $y \in Y$. Es ist $f^{-1}(y) = X \underset{Y}{\times} \text{Spec}(k(y))$, wenn $k(y) = O_y/m_y$ den Restklassenkörper der Faser O_y über $y \in Y$ der Strukturgarbe von Y bezeichnet, bezüglich des kanonischen Morphismus $\text{Spec}(k(y)) \to Y$. $f^{-1}(y)$ ist ein Präschema über $\text{Spec}(k(y))$, also über einem Körper. Unterpräschemata von (X, O_x) sind folgendermaßen definiert: Die abgeschlossenen Unterpräschemata Y entsprechen $1 - 1$ den quasikohärenten Garben $\mathcal{J}$ von Idealen $\subset O_X$: Der Träger Y von $O_X/\mathcal{J}$ ist abgeschlossen in X und liefert den unterliegenden Raum, $O_X\mathcal{J}$ die Strukturgarbe O_Y von Y. So entsprechen die (dann wieder affinen) abgeschlossenen Unterpräschemata des affinen Schemas Spec(A) eindeutig den Idealen $\subset A$. Verschiedene solche können also durchaus denselben unterliegenden Raum haben. Dies ist ein Gesichtspunkt, den W. Gröbner immer schon vertreten hat. Ein offenes Unterpräschema Y von (X, O_X) ist die offene Menge Y versehen mit der Strukturgarbe $O_{X|Y}$. Allgemein ist ein Unterpräschema ein abgeschlossenes Unterpräschema eines offenen Unterpräschemas, der unterliegende Raum also lokal abgeschlossene Teilmenge. Ein Morphismus $f : X \to Y$ heißt separiert, wenn der kanonische

Diagonalmorphismus $X \xrightarrow{\Delta} X \underset{Y}{\times} X$ eine abgeschlossene Einbettung von X
liefert, d.h. wenn X vermöge Δ ein abgeschlossenes Unterpräschema von
$X \underset{Y}{\times} X$ ist. Jedes Präschema X kann, da nur unitäre Ringe betrachtet werden,
als Präschema über $\mathbb{Z}$ aufgefaßt werden: $X \to \text{Spec}(\mathbb{Z})$. X heißt Schema,
wenn dieser kanonische Morphismus separiert ist. Von besonderer Bedeu-
tung sind später Endlichkeitsbedingungen: Ein Schema heißt Noethersch,
wenn es durch endlich viele offene affine Schemata zu Noetherschen Ringen überdeckbar ist, ein Morphismus $f : X \to Y$ von endlichem Typ, wenn
$Y = \bigcup_\alpha V_\alpha$ von offen affinen V_α, so daß $\overline{f}^1(V_\alpha) = \bigcup_{\text{endl.}} U_{\alpha i}$ von endlich vielen offenen affinen $U_{\alpha i}$ zu Ringen $A(U_{\alpha i})$ die endlich erzeugte Algebren über
$A(V_\alpha)$ sind. Ist K ein Körper, so heißt ein Schema K-algebraisch, wenn es von
endlichem Typ über $\text{Spec}(K)$ ist. Die reduzierten, irreduziblen, K-algebrai-
schen Schemata X sind gerade die Chevalleyschen Schemata von lokalen
Ringen in einem Körper. Dabei heißt ein Präschema reduziert, wenn die
lokalen Ringe in seiner Strukturgarbe keine nilpotenten Elemente enthalten.
Man erhält diesen Zusammenhang, indem man vom Körper der rationalen
Funktionen auf X ausgeht, das sind Schnitte in der Strukturgarbe O_X über
offenen in X dichten Teilmengen. Allgemeiner werden für zwei Präschemata
X und Y rationale Abbildungen definiert. Schließlich werden einige schöne
Eigenschaften für quasikohärente Garben auf Präschemata bewiesen (Fortsetzung von quasikohärenten Garben, Fortsetzung von Schnitten in solchen,
Quasikohärenz der direkten Bilder von solchen). Im letzten Paragraphen
werden formelle Schemata eingeführt: Für einen zulässigen Ring A ist $\text{Spf}(A)$
die Menge der offenen Primideale von A. Es ist $\text{Spf}(A) = \text{Spec}(A/J)$ für jedes
Definitionsideal J von A, also eine abgeschlossene Teilmenge von $\text{Spec}(A)$.
Auf $\text{Spf}(A)$ versehen mit der Zariski Topologie erhält man für ein Fundamentalsystem von Nullumgebungen bestehend aus Definitionsidealen J_λ ein pro-
jektives System von Garben von Ringen $O_\lambda = \widetilde{A/J_\lambda}$, wenn $\widetilde{A/J_\lambda}$, die dem Ring
A/J_λ, auf $\text{Spec}(A/J_\lambda) = \text{Spf}(A)$ zugeordnete Garbe von Ringen bezeichnet.
Beschränkt man sich auf eine Basis für die Topologie von $\text{Spf}(A)$ bestehend
etwa aus offenen affinen Teilmengen U, so kann man den projektiven Limes
$O = \varprojlim_{\lambda \in L} O_\lambda$ der Garben O_λ als Garbe von topologischen Ringen auf $\text{Spf}(A)$
betrachten, wenn man für jedes U $\Gamma(U, O) = \varprojlim_{\lambda \in L} \Gamma(U, O_\lambda)$ dem topologischen
projektiven Limes des Systems der diskreten Ringe $\Gamma(U, O_\lambda)$ setzt. $\text{Spf}(A)$
mit dieser Strukturgarbe O von topologischen Ringen ist das formelle affine
Schema $\text{Spf}(A)$ zum zulässigen Ring A. Ein formelles Präschema X ist ganz
analog ein topologischer Raum X mit einer Strukturgarbe O_X von topologischen Ringen, der eine Überdeckung durch offene formelle affine Mengen V_i
besitzt, und ein Morphismus zwischen formellen Präschemata ist ein Morphismus zwischen den topologisch geringten Räumen, der in den Fasern
lokale Homomorphismen induziert. Mit dieser Definition ist die Kategorie
der formalen affinen Schemata dual zur Kategorie der zulässigen Ringe mit

stetigen Ringhomomorphismen als Morphismen. Formelle Schemata treten auf, wenn man ein der Einfachheit halber etwa Noethersches gewöhnliches Präschema längs einer abgeschlossenen Teilmenge X' komplettiert. Dazu sei Φ die Menge der kohärenten Ideale $\mathcal{J}$ mit Träger $(O_X/\mathcal{J}) = X'$. Φ ist nach rechts gefiltert. X' versehen mit der Garbe von topologischen Ringen $\widehat{O} = \varprojlim_{\mathcal{J} \in \Phi} O_X/\mathcal{J}$ ist ein formelles Präschema, die Komplettierung von X längs X', und wird mit $\widehat{X}$ bezeichnet. Ist $X = \mathrm{Spec}(A)$ und X' die dem Ideal $J \subset A$ entsprechende abgeschlossene Teilmenge, so ist X kanonisch isomorph zu $\mathrm{Spf}(\widehat{A})$, wo $\widehat{A}$ die Komplettierung von A nach der J-adischen Topologie ist. Ebenso erhält man für eine kohärente Garbe F auf X eine kohärente Garbe $\widehat{\mathcal{F}} = \varprojlim_{\mathcal{J} \in \Phi}(\mathcal{F} \otimes_{O_X} (O_X/\mathcal{J}))$ auf X. Die Schnitte in $\widehat{O}$ sind nichts anderes als die von O. Zariski eingeführten "holomorphen Funktionen" [Mem. Am. Math. Soc. 5, 90 p. (1951; Zbl 45.24001)]. Vermöge der kanonischen Homomorphismen $O_X \rightarrow O_X/\mathcal{J}$ erhält man einen Homomorphismus der Garben von Ringen $\vartheta : O_X \rightarrow \psi_*(\widehat{O}) = \varprojlim_{\mathcal{J} \in \Phi} O_X/\mathcal{J}$, wenn ψ die Injektion $X' \rightarrow X$ bezeichnet. $i = (\psi, \vartheta) : \widehat{X} \rightarrow X$ ist daher ein Morphismus geringter Räume, und es ist der zu dem aus $\mathcal{F} \rightarrow \mathcal{F} \otimes (O_X/\mathcal{J})$ wie eben erhaltenen Homomorphismus $\mathcal{F} \rightarrow i_*(\widehat{\mathcal{F}})$ adjungierte Homomorphismus $i_*(\mathcal{F}) \xrightarrow{\sim} \widehat{\mathcal{F}}$ ein Isomorphismus. Der Funktor $\mathcal{F} \rightarrow \widehat{\mathcal{F}}$ ist exakt. Ist $f : X \rightarrow Y$ ein Morphismus zwischen Noetherschen Präschemata und sind X', Y' abgeschlossene Teilmengen mit $f(X') \subset Y'$, dann besitzt f eine eindeutig bestimmte Fortsetzung auf die Komplettierungen von X bzw. Y längs X' bzw. Y' $\widehat{f} : \widehat{X} \rightarrow \widehat{Y}$. Schließlich wird die Kategorie der formellen Schemata untersucht und es werden die wichtigsten Definitionen auf formelle Schemata übertragen (Noethersche formelle Präschemata, Morphismen von endlichem Typ, separierte Morphismen usw.).

Das 2. Kapitel behandelt einige neue Klassen von Morphismen, von denen hier im wesentlichen nur funktorielle und globale Eigenschaften betrachtet werden, die oft unschwer aus den Definitionen folgen. Die zu definierenden Klassen von Morphismen ergeben sich aus Klassen von algebraischen Varietäten in der klassischen algebraischen Geometrie nach dem folgenden Prinzip. An die Stelle des dort festen Grundkörpers k oder auch an die Stelle von $\mathrm{Spec}(k)$ tritt hier ein beliebiges Präschema S, und anstatt eine k Varietät V zu betrachten, betrachtet man hier ein Präschema X zusammen mit einem Morphismus $f : X \rightarrow S$. An die Stelle von Eigenschaften von V treten nun Eigenschaften des Morphismus f. Aus diesen allgemeineren "relativen" Eigenschaften erhält man die "absoluten Eigenschaften" im Spezialfall, daß $S = \mathrm{Spec}(k)$ ein Punkt ist. Folgende triviale Tatsache wird hier verallgemeinert: Ist $A \rightarrow B$ ein Homomorphismus zwischen Ringen, so entspricht diesem ein Morphismus zwischen den affinen Schemata $f : \mathrm{Spec}(B) \rightarrow \mathrm{Spec}(A)$. Es ist $A = \Gamma(\mathrm{Spec}(A), \widetilde{A})$, $\widetilde{A}$ die Strukturgarbe von Spec(A). Ganz, analog gilt nun, wenn Spec(A) durch ein beliebiges Präschema S, $\widetilde{A}$ durch dessen Strukturgarbe O_S, ersetzt wird: Ist $\mathcal{B}$ eine quasikohärente (abgekürzt q.k.) Garbe von Ringen auf S und $O_S \to \mathcal{B}$ ein Garbenhomomorphismus, kurz B eine q.k. O_S-Algebra, so gehört dazu ein Schema $X = \operatorname{Spec}(\mathcal{B})$ und ein Morphismus $f : X \to S$. Man erhält rückwärts $\mathcal{B}$ in der Form $\mathcal{B} = f_*(O_X)$. Ein Morphismus f zwischen Präschemata $X \xrightarrow{f} S$ heißt affin wenn $\mathcal{B} = f_*(O_X)$, q.k. ist, $X = \operatorname{Spec}(\mathcal{B})$ und f gleich dem zugehörigen Morphismus $X \to S$ ist. Affine Morphismen sind durch folgende Eigenschaften gekennzeichnet: Es existiert eine Überdeckung $S = \bigcup_\alpha S_\alpha$ von S durch offene affine S_α, so daß die offenen Mengen $f^{-1}(S_\alpha) \subset X$ affin sind. Die Kategorie der "über S affinen Schemata" ist dual zur Kategorie der q.k. O_S-Algebren über S. Für ein beliebiges S Präschema Y und $X = \operatorname{Spec}(\mathcal{B})$ wie oben entsprechen die S-Morphismen $Y \to X = \operatorname{Spec}(\mathcal{B})$ eineindeutig den O_S-Homomorphismen $\mathcal{B} \to f_*(O_Y)$. – Ein vektorielles Faserbündel über S entsteht aus einer q.k. Garbe von O_S-Moduln $\mathcal{E}$ folgendermaßen: Man bildet mit dem O_S-Modul $\mathcal{E}$, die symmetrische Algebra $\mathbf{S}(\mathcal{E})$, eine q.k. O_S-Algebra. $\mathbf{V}(\mathcal{E}) = \operatorname{Spec}(\mathbf{S}(\mathcal{E}))$ ist dann das durch $\mathcal{E}$ definierte vektorielle Faserbündel über S. Die Garbe der Keime von S-Schnitten von $\mathbf{V}(\mathcal{E})$ ist gleich $\check{\mathcal{E}} = \operatorname{Hom}_{O_S}(\mathcal{E}, O_S)$. Die üblichen Vektorraumbündel entsprechen dabei den lokal freien O_S-Moduln $\mathcal{E}$.

Ist B eine graduierte A-Algebra, d.h. eine graduierte Algebra mit einem Ringhomomorphismus $A \to B_0$ (= Bestandteil 0-ten Grades von B), und wird $B_+ = \bigoplus_{n>0} B_n$ von B_1 erzeugt, so wird in bekannter Weise über dem homogenen Spektrum (d.h. der Menge der homogenen Primideale von B, die B_+ nicht umfassen) die Struktur eines Schemas $X = \operatorname{Proj}(B)$ mit einem Morphismus $X \to \operatorname{Spec}(A)$ definiert. Den graduierten B-Moduln entsprechen q.k. O_X-Moduln, insbesondere den graduierten B-Moduln $B(n)$ (Graduierung um n verschoben) invertierbare O_X-Moduln $O_X(n)$ (bezüglich $\otimes$, d.h. lokal zu O_X isomorphe O_X-Moduln). Ist $\varphi : B' \to B$ ein Homomorphismus zwischen graduierten A-Algebren, $X = \operatorname{Proj}(B)$, $X' = \operatorname{Proj}(B')$, so wird durch φ im allgemeinen nur ein (affiner) Morphismus $\Phi = \operatorname{Proj}(\varphi) : G(\varphi) \to X'$ über $\operatorname{Spec}(A)$, eines offenen Unterschemas $G(\varphi)$ von X in X' definiert. Man hat einen kanonischen Homomorphismus $\Phi^*(O_{X'}(n)) \to O_X(n)|_{G(\varphi)}$. Ganz analog erhält man, wenn $\operatorname{Spec}(A)$ durch ein beliebiges Präschema S ersetzt wird: Ist $\mathcal{B}$ eine q.k. graduierte O_S-Algebra, die von $\mathcal{B}_1$ erzeugt wird, so gehört dazu ein Präschema $X = \operatorname{Proj}(\mathcal{B})$ und ein separierter Morphismus $f : X \to S$, ferner invertierbare O_X-Moduln $O_X(n)$. Ebenso gehört zu einem Homomorphismus $\varphi : \mathcal{B}' \to \mathcal{B}$ von q.k. graduierten O_S-Algebren ein Morphismus $\Phi = \operatorname{Proj}(\varphi) : G(\varphi) \to X' = \operatorname{Proj}(\mathcal{B}')$ über S, wo $G(\varphi)$ wieder im allgemeinen nur ein offenes Unterpräschema von $X = \operatorname{Proj}(\mathcal{B})$ ist.

Y sei vermöge $q : Y \to S$ ein Präschema über S, $\mathcal{L}$ ein invertierbarer O_Y Modul. Dann ist $\mathcal{B} = \bigoplus_{n\geq 0} \mathcal{L}^{\otimes n}$ eine q.k. graduierte O_Y-Algebra und $\operatorname{Proj}(\mathcal{B})$ kanonisch isomorph zu Y [entsprechend der Isomorphie $\operatorname{Proj}(A[T]) =$

Spec(A)]. Setzt man nun $q^*(\mathcal{B}') = \mathcal{B}''$, so ist $\mathcal{B}''$ eine q.k. graduierte O_Y-Algebra zu der das Y-Schema $\mathrm{Proj}(q^*(\mathcal{B}')) = \mathrm{Proj}(\mathcal{B}') \underset{S}{\to} \times Y = X' \underset{S}{\to} \times Y$ gehört. Den Homomorphismen $\varphi : \mathcal{B}'' \to \mathcal{B}$ entsprechen nun wieder Morphismen

$$\Phi = \mathrm{Proj}(\varphi) : G(\varphi) \to X' \underset{S}{\times} Y \xrightarrow{\mathrm{pr}_{X'}} X', \quad G(\varphi) \text{ offen in } Y.$$

Ein projektives Faserbündel X' über S entsteht aus einem q.k. O_Y-Modul $\mathcal{E}$ folgendermaßen: Man bildet die symmetrische Algebra $S(\mathcal{E}) = \mathcal{B}'$, eine q.k. O_S-Algebra, und setzt $\mathrm{Proj}(S(\mathcal{E})) = \mathbf{P}(\mathcal{E}) = X'$. Ein Homomorphismus $\psi : q^*(\mathcal{E}) \to \mathcal{L}$ induziert einen Homomorphismus der Algebren $q^*(\mathbf{S}(\mathcal{E})) \to \mathcal{B}$, also von $\mathcal{B}'' = q^*(\mathcal{B}') \to \mathcal{B}$. So erhält man eine (1–1)-Beziehung zwischen S-Morphismen $\Phi : Y \to \mathbf{P}(\mathcal{E}) = X'$ von ganz Y in X' und Klassen von Paaren $(\mathcal{L}, \psi)$, bestehend aus einem invertierbaren O_Y-Modul $\mathcal{L}$ und einem surjektiven Homomorphismus $\psi : q^*(\mathcal{E}) \to \mathcal{L}$. Es ist $\mathcal{L} \cong \Phi^*(O_{X'}(1))$. Insbesondere entsprechen die S-Schnitte von $\mathbf{P}(\mathcal{E})$ 1 – 1 den q.k. O_S-Untermoduln $\mathcal{F}$ von $\mathcal{E}$ mit $\mathcal{E}/\mathcal{F}$ invertierbar. Dies führt zu folgenden Definitionen von aus der klassischen algebraischen Geometrie bekannten Begriffen jetzt in der allgemeinen Theorie der Schemata: Es sei $Y \xrightarrow{q} S$ wie oben ein S-Präschema. Ein invertierbarer O_Y-Modul $\mathcal{L}$ heißt streng vollständig (très ample) bezüglich q oder bezüglich S, wenn ein q.k. O_S-Modul $\mathcal{E}$, und eine S-Einbettung $\Phi : Y \to \mathbf{P}(\mathcal{E}) = X'$ existiert, so daß $\mathcal{L} \cong \Phi^*(O_{X'}(1))$ ist. Y ist also dann vermöge Φ ein S-Unterschema von $\mathbf{P}(\mathcal{E})$. Ist q quasikompakt, das heißt ist das Urbild jeder offenen quasikompakten Menge in S quasikompakt in Y, so ist $\mathcal{L}$ genau dann streng vollständig bezüglich q, wenn $q_*(\mathcal{L})$ q.k., der kanonische Homomorphismus $q^*(q_*(\mathcal{L})) \to \mathcal{L}$ surjektiv und der zugehörige Morphismus $\Phi : Y \to \mathbf{P}(q_*(\mathcal{L}))$ eine Einbettung ist. Ist S Noethersch und q von endlichem Typ, so existiert ein kohärenter O_S-Modul $\mathcal{E}$ mit den obigen Eigenschaften bezüglich $\mathcal{L}$. X sei ein quasikompaktes Schema, $\mathcal{L}$ ein invertierbarer O_X-Modul. Man kann X als $\mathbf{Z}$-Schema auffassen: $X \xrightarrow{p} \mathrm{Spec}(\mathbf{Z})$. Mit $B = \bigoplus_{n \geq 0} \Gamma(X, \mathcal{L}^{\otimes n})$ hat man einen kanonischen Homomorphismus der graduierten O_X-Algebren $\varphi : p^*(\widetilde{B}) \to \bigoplus_{n \geq 0} \mathcal{L}^{\otimes n} \cdot \mathcal{L}$ heißt vollständig, wenn für den zugehörigen Morphismus $\Phi : X \supset G(\varphi) \to \mathrm{Proj}(B)$ gilt: $G(\varphi) = X$ und Φ ist eine offene und dichte Einbettung von X in $\mathrm{Proj}(B)$. Bezeichnet man die Garbe $\mathcal{F} \otimes \mathcal{L}^{\otimes n}$ mit $\mathcal{F}(n)$, so ist dies äquivalent zu

(i) Ist $\mathcal{F}$ q.k. und endlich erzeugter O_X-Modul (de type fini), so wird $\mathcal{F}(n)$ für alle $n \geq n_0 > 0$ durch seine Schnitte über X erzeugt, und zu

(ii) Ist $\mathcal{F}$ q.k. und endlich erzeugter O_X-Modul, so existieren $n > 0, k > 0$, so daß $\mathcal{F}$ Quotient des O_X-Moduls $\mathcal{L}^{\otimes (-n)} \otimes O_X^k$ ist.

Ist S affin, $q : X \to S$ separiert und quasi-kompakt, $\mathcal{L}$ ein invertierbarer O_X-Modul, dann sind äquivalent

(i) $\mathcal{L}$ ist vollständig;

(ii) Es existiert ein $n_0 > 0$, so daß für $n \geq n_0 \mathcal{L}^{\otimes n}$ streng vollständig bezüglich q ist;

(iii) Es existiert ein $n > 0$, so daß $\mathcal{L}^{\otimes n}$ streng vollständig bezüglich q ist.

Ist $f : X \to S$ quasikompakt, $\mathcal{L}$ ein invertierbarer O_X-Modul und $\mathcal{B} = \bigoplus_{n \geq 0} f_*(\mathcal{L}^{\otimes n})$, dann sind äquivalent

(i) Es existiert eine Überdeckung $S = \bigcup_\alpha U_\alpha$ durch offne affine U_α, so daß $\mathcal{L}|_{X_\alpha}$ über $X_\alpha = f^{-1}(U_\alpha)$ vollstndig ist;

(ii) $\mathcal{B}$ ist q.k. und für den dem kanonischen Homomorphismus $\varphi : f^*(\mathcal{B}) \to \bigoplus_{n \geq 0} \mathcal{L}^{\otimes n}$ entsprechenden S-Morphismus $\Phi : X \supset G(\varphi) \to \text{Proj}(\mathcal{B}) = P$ gilt: $G(\varphi) = X$, und Φ ist eine offene und dichte Einbettung von X in P.

$\mathcal{L}$ heißt dann relativ vollständig bezüglich f, kurz f-vollständig oder S-vollständig. Ist S affin, so ist $\mathcal{L}$ S-vollständig genau dann, wenn $\mathcal{L}$ vollständig ist.

Ist X ein quasikompaktes Schema und f ein separierter, quasikompakter Morphismus $f : X \to S$, so sind äquivalent

(i) $\mathcal{L}$ ist f vollständig;

(ii) Für jeden q.k. endlich erzeugten O_X-Modul $\mathcal{F}$ existiert ein $n_0 > 0$, so daß für $n \geq n_0$ der kanonische Homomorphismus $f^*(f_*(\mathcal{F} \otimes \mathcal{L}^{\otimes n})) \to \mathcal{F} \otimes \mathcal{L}^{\otimes n}$ surjektiv ist;

(iii) Für jedes q.k. endlich erzeugte O_X-Ideal $\mathcal{J}$ existiert ein $n > 0$, so daß der kanonische Homomorphismus $f^*(f_*(\mathcal{J} \otimes \mathcal{L}^{\otimes n})) \to \mathcal{J} \otimes \mathcal{L}^{\otimes n}$ surjektiv ist.

Ist S quasikompakt, $f : X \to S$ von endlichem Typ, dann sind äquivalent

(i) $\mathcal{L}$ ist f-vollständig;

(ii) Es existiert ein $n_0 > 0$, so daß für alle $n \geq n_0$ $\mathcal{L}^{\otimes n}$ streng vollständig bezüglich f ist;

(iii) Es existiert ein $n > 0$, so daß $\mathcal{L}^{\otimes n}$ streng vollständig bezüglich f ist.

Ist S ein quasikompaktes Präschema, $\mathcal{B}$ eine q.k. endlich erzeugte O_S-Algebra, $X = \text{Proj}(\mathcal{B})$ und $f : X \to S$ der kanonische Morphismus, dann ist f von endlichem Typ und es existiert ein $d > 0$, so daß $O_X(d)$ invertierbar und f-vollständig ist.

§5 enthält das wichtige Serresche Kriterium: Ist X ein quasikompaktes Schema oder ein Präschema mit Noetherschem unterliegendem Raum, so ist X genau dann affin, wenn der Funktor $\mathcal{F} \to \Gamma(X, \mathcal{F})$ auf der Kategorie der q.k. O_X-Moduln $\mathcal{F}$ exakt ist.

Ebenso ist der separierte quasikompakte Morphismus $f : X \to Y$ genau dann affin, wenn f_* exakt auf der Kategorie der q.k. O_X-Moduln ist. Ein Morphismus $f : X \to Y$ heißt quasiprojektiv kurz q.p. oder X ist q.p. über Y, wenn f von endlichem Typ ist, und wenn ein f-vollständiger O_X-Modul existiert. Für ein quasikompaktes Y ist dies äquivalent mit der Existenz eines q.k. endlich erzeugten O_Y-Moduls $\mathcal{E}$, so daß X Y-isomorph einem Unterpräschema von $\mathbb{P}(\mathcal{E})$ ist. Existiert ein vollständiger O_Y-Modul $\mathcal{L}$, so kann $\mathcal{E}$ sogar $= O_Y^n$ gewählt werden.

Ein Morphismus $f : X \to Y$ heißt eigentlich (propre), wenn f separiert und von endlichem Typ ist und wenn für jedes Präschema Y' über Y die Projektion $X \underset{Y}{\times} Y' \to Y'$ ein abgeschlossener Morphismus ist (das heißt das Bild einer abgeschlossenen Menge wieder abgeschlossen ist).

Ein Morphismus $X \to Y$ heißt projektiv, wenn X Y-isomorph einem abgeschlossenen Unterpräschema eines projektiven Faserbündels $\mathbb{P}(\mathcal{E})$ zu einem endlich erzeugten q.k. O_Y-Modul $\mathcal{E}$ ist.

Ist Y ein quasikompaktes Schema, so ist $X \xrightarrow{f} Y$ projektiv genau dann, wenn es quasiprojektiv und eigentlich ist.

Schließlich wird das Chowsche Lemma in folgender Form bewiesen: S sei Noethersch oder es sei S ein quasikompaktes Schema und X habe nur endlich viele irreduzible Komponenten, $X \to S$ sei von endlichem Typ. Dann existiert ein q.p. S-Schema X' und ein projektiver und surjektiver S-Morphismus $f : X' \to X$ und ein offenes $U \subset X$ mit $U' = \bar{f}^1(U)$ dicht in X' und $f|_{U'} : U' \to U$ ein Isomorphismus $X \to S$ ist genau dann eigentlich, wenn $X' \to S$ sogar projektiv ist, und U kann dann sogar dicht in X gewählt werden. Ein affiner Morphismus $f : X \to S$ heißt ganz bzw. endlich, wenn es eine Überdeckung $S = \bigcup_\alpha S_\alpha$ durch offene affine S_α zu Ringen A_α gibt, so daß die $\bar{f}^1(S_\alpha)$ affin zu Ringen B_α sind, die ganz bzw. endliche Moduln über A_α sind. Ist $X \to S$ ganz und von endlichem Typ, so ist $X \to S$ endlich. Ein endlicher Morphismus ist projektiv.

Es werden quasiendliche Morphismen sowie die ganze Abschließung eines Präschemas definiert.

Für endliche Morphismen $X' \to X$ und invertierbare $O_{X'}$-Moduln $\mathcal{L}$ wird $N_{X'|X}(\mathcal{L})$, ein invertierbarer O_X-Modul, definiert, und damit werden Kriterien für relative Vollständigkeit von $\mathcal{L}$ gewonnen. Ferner wird der folgende Satz von Chevalley bewiesen. Ist X ein affines Schema und Y ein Noethersches Präschema, $f : X \to Y$ ein endlicher und surjektiver Morphismus, so ist Y affin.

In §7 werden bewertungstheoretische Kriterien für die Eigenschaften eines Morphismus, separiert bzw. eigentlich zu sein, angegeben, die den bekannten Kriterien im Funktionenkörper im Falle irreduzibler algebraischer Varietäten entsprechen. Der Fall der algebraischen Kurven wird eingeordnet.

In §8 schließlich wird der Prozeß des Aufblasens, der sich mit den zuvor definierten Begriffen besonders einfach formulieren läßt, behandelt: Ist Y ein Präschema (bzw. ein lokal integeres Präschema, das heißt lokal irreduzibel und reduziert) und $\mathcal{J}$ ein q.k. Ideal $\subset O_Y$ (bzw. ein "gebrochenes" q.k. Ideal $\subset \mathcal{R}(Y)$, der Garbe der Keime rationaler Funktionen auf Y), dann ist das Y-Schema $X = \mathrm{Proj}\left(\bigoplus_{n \geq 0} \mathcal{J}^n\right)$ das durch Aufblasen von J aus Y erhaltene Präschema. Schließlich wird der projizierende Kegel und die projektive

Abschließung in der Sprache der Schemata behandelt und das folgende Kriterium von Grauert für Vollständigkeit bewiesen: Y sei ein Präschema, $p : X \to Y$ ein separierter quasikompakter Morphismus, $\mathcal{L}$ ein invertierbarer O_X-Modul. $\mathcal{L}$ ist genau dann p vollständig, wenn ein Y-Präschema C und ein Y-Schnitt $\varepsilon : Y \to C$ von C existiert, sowie ein Y Morphismus $q : \mathbf{V}(\mathcal{L}) \to C$, so daß

$$
\begin{array}{ccc}
X & \xrightarrow{\ j\ } & \mathbf{V}(\mathcal{L}) \\
{\scriptstyle p}\downarrow & & \downarrow{\scriptstyle q} \\
Y & \xrightarrow{\ \varepsilon\ } & C
\end{array}
\tag{i}
$$

wo j der Nullschnitt des Geradenbündels $\mathbf{V}(\mathcal{L})$ ist, kommutativ ist, und (ii) die Einschränkung von q auf $\mathbf{V}(\mathcal{L}) - j(X)$ eine offene quasikompakte Einbettung $\mathbf{V}(\mathcal{L}) - j(X) \to C$ ist, deren Bild $\varepsilon(Y)$ nicht trifft.

Dem ersten Teil des 3. Kapitels vorangestellt sind wieder fünf Paragraphen des Kapitels 0. Einer derselben befaßt sich zunächst mit repräsentierbaren Funktoren. Ist $\mathbf{C}$ irgendeine Kategorie und $\mathbf{E}$ die Kategorie der Mengen, so wird für ein festes Objekt $X \in \mathbf{C}$ durch $Y \rightsquigarrow h_X(Y) = \mathrm{Hom}_{\mathbf{C}}(Y, X)$ ein Kontrafunktor von $\mathbf{C}$ in $\mathbf{E}$ definiert. Man erhält so, wenn X die Kategorie $\mathbf{C}$ durchläuft einen Funktor $h : \mathbf{C} \to \mathbf{Hom}(\mathbf{C}^0, \mathbf{E})$ von $\mathbf{C}$ in die Kategorie der Kontrafunktoren von $\mathbf{C}$ in $\mathbf{E}$. Ist F irgendein Kontrafunktor $\in \mathbf{Hom}(\mathbf{C}^0, \mathbf{E})$, so hat man eine natürliche Bijektion von $\mathrm{Hom}(h_X, F)$ in $F(X)$. Für $F = h_X$, erhält man so: Der Funktor $h : \mathbf{C} \to \mathbf{Hom}(\mathbf{C}^0, \mathbf{E})$ ist vollkommen treu (pleinement fidèle), d.h. die kanonische Abbildung $\mathrm{Hom}(X, X') \to \mathrm{Hom}(h_X, h_{X'})$ ist eine Bijektion, und definiert eine Äquivalenz von $\mathbf{C}$ mit der vollen Unterkategorie der Funktoren der Form h_X, $X \in \mathbf{C}$. Diese Kontrafunktoren heißen repräsentierbare Kontrafunktoren.

Eine innere Verknüpfung (loi de composition interne) auf X wird nun definiert durch einen Morphismus von Funktoren $\gamma_X : h_X \times h_X \to h_X$. Dabei ist für alle $Y \in \mathbf{C}$ $(h_X \times h_X)(Y) = h_X(Y) \times h_X(Y)$, wo das rechte $\times$ das übliche kartesische Produkt in $\mathbf{E}$ bezeichnet. Für jedes $Y \in \mathbf{C}$ ist demnach $\gamma_X(Y)$ eine Abbildung $\gamma_X(Y) : h_X(Y) \times h_X(Y) \to h_X(Y)$, also eine innere Verknüpfung in $h_X(Y)$, so daß für jeden Morphismus $u : Y \to Y'$ in $\mathbf{C}$

$$
\begin{array}{ccc}
h_X(Y') \times h_X(Y') & \xrightarrow{h_X(u) \times h_X(u)} & h_X(Y) \times h_X(Y) \\
{\scriptstyle \gamma_X(Y')}\downarrow & & \downarrow{\scriptstyle \gamma_X(Y)} \\
h_X(Y') & \xrightarrow[h_X(u)]{} & h_X(Y)
\end{array}
$$

kommutativ, d.h. $h_X(u)$ ein Homomorphismus von $h_X(Y')$ in $h_X(Y)$ ist. Wenn γ_X so beschaffen ist, daß $\gamma_X(Y)$ für alle $Y \subset \mathbf{C}$ eine Gruppe auf $h_X(Y)$ definiert, so heißt X mit γ_X eine $\mathbf{C}$-Gruppe oder eine Gruppe in $\mathbf{C}$. Analog definiert man $\mathbf{C}$-Ringe usw. Existiert in $\mathbf{C}$ das Produkt $X \times X$, dann ist $h_{X \times X} \times h_X \times h_X$ und der funktorielle Morphismus γ_X definiert eindeutig einen Morphismus

$c_X \in \mathrm{Hom}_C(X \times X, X)$, und eine innere Verknüpfung auf X wird daher auch gegeben durch einen Morphismus $c_X : X \times X$.

Ein weiterer Paragraph bringt allgemeines über die von C. Chevalley eingeführten konstruierbaren Mengen, insbesondere in Noetherschen Räumen. Konstruierbare Mengen sind im Falle der klassischen algebraischen Geometrie die endlichen Durchschnitte von in der Zariski-Topologie abgeschlossenen und offenen Teilmengen. Schließlich bringt ein Paragraph weitere Eigenschaften flacher Moduln (modules plats) und Erweiterungen, bezüglich deren Beweis auf Bourbaki, Éléments de mathématique. Fasc. XXVIII: Algèbre commutative. Chap. III., IV. (Paris 1961) verwiesen wird, und ein Paragraph Ergänzungen hauptsächlich zu des Verf. Tôhoku-Arbeit [Tohoku Math. J., II. Ser. 9, 119–221 (1957; Zbl 0118.26104)] über spektrale Folgen, Hypercohomologie eines Funktors bezüglich eines Komplexes, Übergang zu induktiven Limites in der Hypercohomolagie und anderes. Ferner werden Ergänzungen zur Garbencohomologie, die höheren direkten Bilder, das sind die rechts abgeleiteten Funktoren des linksexakten Funktors "direktes Bild" und die lokalen Ext-Funktoren sowie die Hypercohomologie des Funktors "direktes Bild" behandelt. Schließlich ist ein Paragraph noch den projektiven Limites gewidmet, insbesondere wird untersucht, wann der Funktor $\varprojlim$ exakt ist (Bedingung von Mittag-Leffler).

Von dem eigentlichen 3. Kapitel, von dem in einer Übersicht die Titel von 11 Paragraphen angegeben werden, enthält dieser 1. Teil die Paragraphen 1–5. Inzwischen ist ein zweiter Teil mit den Paragraphen 6 und 7 erschienen. Von den weiteren Paragraphen liegt zur Zeit z.B. schon teilweise §9, Relative und lokale Cohomologie, lokale Dualität, als Seminarausarbeitung des Séminaire de Géoétrie algébrique 1962 am Inst. haut. Étud. sci. in einer vorläufigen Form vor. Die ersten 5 Paragraphen des ersten Teils enthalten in den Paragraphen 1 und 2 die Serreschen Resultate aus FAC über die Cohomologie affiner und projektiver Varietäten in dem allgemeinen Rahmen der Schemata sowie in den folgenden 3 Paragraphen des Verf. Erweiterung des Endlichkeitssatzes auf eigentliche Morphismen und auf formelle Schemata sowie die fundamentalen Sätze über die Beziehungen zwischen gewöhnlicher und formeller Cohomologie zusammen mit einem Existenzsatz, die z.B. einen natürlichen Beweis des Zariskischen Zusammenhangssatzes liefern.

Im Einzelnen beginnt §1 mit einigen Eigenschaften des Koszul-Komplexes (complexe de l'algèbre extérieure) des Ringes A zu einem Elementsystem $\mathbf{f} = (f_1, \ldots, f_r)$, der hier mit $K_\bullet(\mathbf{f})$ bezeichnet wird. Für ein zweites Elementsystem $\mathbf{g} = (g_1, \ldots, g_r)$ definiert man $\mathbf{fg} = (f_1 g_1, f_2 g_2, \ldots, f_r g_r)$ und hat einen kanonischen Homomorphismus $\varphi_g : K_\bullet(fg) \to K_\bullet(f)$. Für einen A-Modul M sei $K_\bullet(\mathbf{f}, M) = K_\bullet(\mathbf{f}) \underset{A}{\to} \times M$, $K^\bullet(\mathbf{f}, M) = \mathrm{Hom}_A(K_\bullet(\mathbf{f}), M)$ und $H_\bullet(\mathbf{f}, M) = H_\bullet(K_\bullet(\mathbf{f}, M))$ die Homologie bzw. $H^\bullet(\mathbf{f}, M) = H^\bullet(K^\bullet(\mathbf{f}, M))$ die Cohomologie der Komplexe. Ist das Ideal $(\mathbf{f}) = A$, dann sind die Komplexe $K_\bullet(\mathbf{f})$ und $K^\bullet(\mathbf{f})$ homotop trivial, also ist $H_\bullet(\mathbf{f}, M) = H^\bullet(\mathbf{f}, M) = 0$. Die Bedeutung des Koszul-Komplexes für die Cohomologie liegt in folgender Tatsache: Ist X ein Präschema mit Noetherschem unterliegendem Raum oder ein quasikompaktes Schema, $\mathcal{F}$ ein q.k. (d.h. quasikohärenter) O_X-Modul, $A = \Gamma(X, O_X)$, $M = \Gamma(X, \mathcal{F})$, $\mathbf{f} = (f_1, \ldots, f_r)$ ein System von Elementen aus A, $U_i = X_{f_i}$ die offene Menge der $x \in X$ mit $f_i(x) \neq 0$, $U = \bigcup_{i=1}^{r} U_i$, $\mathfrak{U}$ die Überdeckung $(U_i)_{1 \leq i \leq r}$ von U, so hat man einen kanonischen Isomorphismus der Gruppe der alternierenden Čechschen p-Koketten zur Überdeckung $\mathfrak{U}$ mit Koeffizienten in $\mathcal{F}$

$$C^p(\mathfrak{U}, \mathcal{F}) \xrightarrow{\sim} C_{p+1}\big((\mathbf{f}), M\big) \underset{\mathrm{def}}{=} \varinjlim_n K^{p+1}(\mathbf{f}^n, M),$$

wo rechts der induktive Limes des Systems von $(p + 1)$-Bestandteilen der KoszulKomplexe $K^{\bullet}(\mathbf{f}^n, M)$ zu den Elementsystemen $(\mathbf{f}^n)$ mit den Homomorphismen $\varphi_{\mathbf{f}^{m-n}} : K^{\bullet}(\mathbf{f}^n, M) \to K^{\bullet}(\mathbf{f}^m, M)$ für $0 \leq n \leq m$ steht. Dieser Isomorphismus ist mit Korandbildung verträglich. Somit hat man, wenn

$$C^{\bullet}\big((\mathbf{f}), M\big) \underset{\mathrm{def}}{=} \varinjlim_n K^{\bullet}(\mathbf{f}^n, M)$$

und

$$H^q\big((\mathbf{f}), M\big) \underset{\mathrm{def}}{=} H^q\big(C^{\bullet}(\mathbf{f}), M\big) = \varinjlim_n H^q(\mathbf{f}^n, M)$$

kanonische Isomorphismen der Čechschen Cohomologiegruppen zur Überdeckung $\mathfrak{U}$ mit Koeffizienten in $\mathcal{F}$, $H^p(\mathfrak{U}, \mathcal{F}) \xrightarrow{\sim} H^{p+l}((\mathbf{f}), M)$ für $p \geq 1$ und eine kanonische exakte Folge

$$0 \to H^0\big((\mathbf{f}), M\big) \to M \to H^0(\mathfrak{U}, \mathcal{F}) \to H^1\big((\mathbf{f}), M\big) \to 0.$$

Ist nun insbesondere X ein affines Schema, so hat man beliebig feine endliche Überdeckungen von X durch Mengen $X_{f_i} = D(f_i)$, $f_i \in A$, und das von einem zugehörigen Elementsystem erzeugte Ideal $(f_i)_{1 \leq i \leq r} = A$. Übergang zum Limes liefert für die Čechsche Cohomologie [vgl. R. *Godement*, Topologie algébrique et théorie des faisceaux. Paris: Hermann & Cie. (1958; Zbl 80.16201)] $\check{H}^p(X, \mathcal{F}) = 0$ und ebenso für jedes $X_f \check{H}^p(X_f, \mathcal{F}) = 0$ für $p > 0$. Dann ist aber auch die Garbencohomologie $H^p(X, \mathcal{F}) = 0$ für $p > 0$. Als Corollar ergibt sich. Ist $f : X \to Y$ affin, $\mathcal{F}$ ein q.k. O_X-Modul, so ist $R^q f_*(\mathcal{F}) = 0$ für $q > 0$ und man hat kanonische Isomorphismen $H^p(Y, f_*(\mathcal{F})) \xrightarrow{\sim} H^p(X, \mathcal{F})$ für jedes p. Für ein beliebiges Schema X und eine Überdeckung $\mathfrak{U} = (U_\alpha)$ von X durch offene affine Mengen U_α und jeden q.k. O_X-Modul $\mathcal{F}$ hat man daher nach Leray $H^{\bullet}(X, \mathcal{F}) \cong \check{H}^{\bullet}(\mathfrak{U}, \mathcal{F})$.– §2 enthält die Cohomologietheorie projektiver Morphismen. Analog zu obigem Ergebnis über die Čech Cohomologie zu einer bestimmten Überdeckung $\mathfrak{U} = (U_i)_{1 \leq i \leq r}$ von $U = \bigcup_i U_i$ mit Koeffizienten in einem q.k. O_X-Modul $\mathcal{F}$ erhält man unter denselben Voraussetzungen über

X für einen invertierbaren O_X-Modul $\mathcal{L}$ mit $S = \Gamma_*(X, \mathcal{L}) \underset{\text{def}}{=} \bigoplus_{n \in \mathbb{Z}} \Gamma(X, \mathcal{L}^{\otimes n})$,

$$M = \Gamma_*(\mathcal{L}, \mathcal{F}) \underset{\text{def}}{=} \bigoplus_{n \in \mathbb{Z}} \Gamma(X, \mathcal{F} \oplus \mathcal{L}^{\otimes n})$$

und ein System $(\mathbf{f}) = (f_1, \ldots, f_r)$ von homogenen Elementen aus S mit analog definierten $U_i = X_{f_i}$, für die graduierten S-Moduln

$$H^p\big(\mathfrak{U}, \mathcal{F}(*)\big) \underset{\text{def}}{=} \bigoplus_{n \in \mathbb{Z}} H^p\big(\mathfrak{U}, \mathcal{F}(n)\big) \overset{\sim}{\to} H^{p+1}\big((\mathbf{f}), M\big) \text{ für } p \geq 1 \quad \text{und}$$

$$0 \to H^0\big((\mathbf{f}), M\big) \to M \to H^0\big(\mathfrak{U}, \mathcal{F}(*)\big) \to H^1\big((\mathbf{f}), M\big) \to 0$$

exakt. Ist X ein quasikompaktes Schema und sind die $U_i = X_{f_i}$ affin, so ist nach Leray $H^p(\mathfrak{U}, \mathcal{F}(*)) \cong H^p(U, \mathcal{F}(*))$. Man hat damit also eine Möglichkeit gewonnen, $H^p(U, \mathcal{F}(*))$ zu berechnen. Analog gilt für einen positiv graduierten Ring S, einen graduierten S-Modul M und homogene Elemente f_i ($1 \leq i \leq r$) $\in S_+$, wenn $X = \text{Proj}(S)$, $U_i = D_+(f_i)$, $U = \bigcup_i U_i$, $\widetilde{M(n)}$ den zu dem S-Modul $M(n)$ gehörigen q.k. O_X-Modul bezeichnet: $H^p(U, \widetilde{M(*)}) \overset{\sim}{\to} H^{p+1}((\mathbf{f}), M)$ und

$$0 \to H^0\big((\mathbf{f}), M\big) \to M \to H^0\big(U, \widetilde{M(*)}\big) \to H^1\big((\mathbf{f}), M\big) \to 0$$

ist exakt. Dies erlaubt im Falle $S = A[T_0, \ldots, T_r]$, $(\mathbf{f}) = (T_0, \ldots, T_r)$ die explizite Berechnung der $H^i(X, O_X(*))$ für $X = \text{Proj}(S)$. Damit ist man nun in der Lage, das Fundamentaltheorem von Serre für projektive Morphismen durch Reduktion auf den Fall $X = \text{Proj}(S)$, $S = A[T_0, \ldots, T_r]$, und in diesem Fall durch absteigende Induktion nach q zu beweisen: Ist Y ein Noethersches Präschema, $f : X \to Y$ ein eigentlicher Morphismus, $\mathcal{L}$ ein invertierbarer bezüglich f vollständiger O_X-Modul, dann gilt für jeden kohärenten O_X-Modul $\mathcal{F}$ mit $\mathcal{F}(n) = \mathcal{F} \underset{O_Y}{\times} \mathcal{L}^{\otimes n}$.

(i) Die $R^q f_*(\mathcal{F})$ sind kohärente O_Y-Moduln.

(ii) Es existiert ein n_0, so daß für $n \geq n_0$ $R^q f_*(\mathcal{F}(n)) = 0$ für alle $q > 0$.

(iii) Es existiert ein n_0, so daß für $n \geq n_0$ der kanonische Homomorphismus $f^*(f_*(\mathcal{F}(n))) \to \mathcal{F}(n)$ surjektiv ist.

Umgekehrt folgt für einen beliebigen invertierbaren O_X-Modul $\mathcal{L}$ aus (ii) die Vollständigkeit von $\mathcal{L}$ bezüglich f. Als Anwendung ergibt sich zunächst wie bei Serre für ein Noethersches Präschema Y, eine positiv graduierte q.k. endlich erzeugte O_Y-Algebra $\mathcal{B}$ und $X = \text{Proj}(\mathcal{B})$ eine Äquivalenz zwischen der Kategorie der kohärenten O_X-Moduln und einer Quotientenkategorie einer Unterkategorie der Kategorie der q.k. graduierten $\mathcal{B}$-Moduln über Y. (Bemerkung des Referenten: Eine entsprechende Äquivalenz erhält man für die Kategorie der q.k. O_X-Moduln.) Ferner erhält man folgenden Satz: Ist Y ein Noethersches, integeres Präschema, X integer und $f : X \to Y$ ein

94

projektiver birationaler Morphismus; dann existiert ein kohärentes gebrochenes Ideal $\mathcal{J} \subset \mathcal{R}(Y)$, $\mathcal{R}(Y)$ die Garbe der Keime rationaler Funktionen, auf Y, so daß X Y-isomorph ist dem durch Aufblasen von $\mathcal{J}$ aus Y erhaltenen Präschema. Insbesondere gilt: Sind X und Y zwei integere über k projektive Schemata und ist $f : X \to Y$ ein birationaler k Morphismus, dann ist X k-isomorph zu einem Y-Schema, das durch Aufblasen eines abgeschlossenen (nicht notwendig reduzierten) Unterschemas Y' von Y aus Y entstanden ist.
§3 enthält den Endlichkeitssatz für eigentliche Morphismen: Ist Y ein lokal Noethersches Präschema, $f : X \to Y$ ein eigentlicher Morphismus, dann sind für jeden kohärenten O_X-Modul $\mathcal{F}$ die höheren Bilder $R^q f_*(\mathcal{F})$ kohärente O_Y-Moduln für $q \geq 0$. Der Beweis verwendet das "Lemma de dévissage". X sei ein Noethersches Präschema, $\mathbf{K}$ die abelsche Kategorie der kohärenten O_X-Moduln, $\mathbf{K}'$ eine exakte Untermenge von $\mathbf{K}$ (d.h. $0 \in \mathbf{K}'$ und $0 \to A' \to A \to A'' \to 0$ exakt in $\mathbf{K}$, zwei der $A', A, A'' \in \mathbf{K}'$, so folgt auch das dritte $\in \mathbf{K}'$), X' eine abgeschlossene Teilmenge des Raumes X. Existiert nun für jede abgeschlossene irreduzible Teilmenge Y von X' mit allgemeinem Punkt y ein O_X-Modul $\mathcal{G} \in \mathbf{K}'$, so daß die Faser $\mathcal{G}_y$ ein $k(y)$ Vektorraum der Dimension 1 ist, so enthält $\mathbf{K}'$ jeden kohärenten O_X-Modul mit Träger in X'.
Ein entsprechender Endlichkeitssatz wird für formelle Schemata bewiesen.
§4 schließlich bringt die fundamentalen Tatsachen über Vertauschbarkeit von direktem Bild und Komplettierung bei eigentlichen Morphismen: X, Y seien Noethersehe Präschemata, $f : X \to Y$ ein eigentlicher Morphismus, $Y' \subset Y$ abgeschlossen und $X' = f^{-1}(Y') \subset X$, $\widehat{X}$ bzw. $\widehat{Y}$ die durch Komplettierung von X bzw. Y längs X' bzw. Y' erhaltenen formellen Präschemata, $\widehat{f} : \widehat{X} \to \widehat{Y}$ die Komplettierung von f, F ein kohärenter O_X-Modul und $\widehat{F}$ seine Komplettierung im Sinne von Kap. I, ein kohärenter $O_{\widehat{X}}$-Modul, $\mathcal{J}$ ein q.k. Ideal $\subset O_Y$ mit Träger $(O_Y/\mathcal{J}) = Y'$. Bezeichnet man für $k > 0$ die kohärenten O_X-Moduln $\mathcal{F} \otimes_{O_Y}(O_Y/\mathcal{J}^{k+1})$ mit $\mathcal{F}_k$, so hat man ein kommutatives Diagramm von kanonischen topologischen Isomorphismen

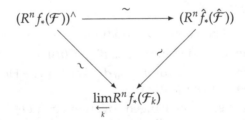

und $R^n \widehat{f_*}(\widehat{\mathcal{F}})$ ist ein kohärenter $O_{\widehat{X}}$-Modul.
Als Folgerung erhält man: Ist Y lokal Noethersch, $f : X \to Y$ eigentlich und F kohärenter O_X-Modul, dann ist für $y \in Y$ die Faser $(R^n f_*(\mathcal{F}))_y$ ein endlich erzeugter O_y-Modul, also separiert in der m_y-adischen Topologie und man hat einen kanonischen topologischen Isomorphismus

$$\left(R^n f_*(\mathcal{F})\right)_y^{\wedge} \xrightarrow{\sim} \varprojlim_k H^n\left(\bar{f}^1(y), \mathcal{F} \bigoplus_{O_Y}(O_y/m_y^k)\right).$$

Für $n = 0$ erhält man so den Zusammenhangssatz von Zariski: Ist Y ein lokal Noethersches Präschema, $f : X \to Y$ eigentlich, so ist $f_*(O_X)$ eine kohärente O_Y-Algebra. Y' sei daszu dieser gehörige über Y endliche affine Y-Schema, also mit $f_*(O_{Y'}) = f_*(O_X)$. Dann hat man (vgl. Kap. II) eine kanonische Faktorisierung von f in $X \xrightarrow{f'} Y' \to Y$ (Steinsche Faktorisierung). f' ist eigentlich, $f'_*(O_X) \simeq O_Y$, und die Faser $\overline{f'}^{-1}(y)$ von f' sind zusammenhängend und nicht leer für alle $y' \in Y'$. Daraus ergeben sich mannigfache Folgerungen unter anderem Zariskis "main theorem". – Ein Existenztheorem in §5 vervollständigt diese Theorie: A sei ein adischer Noetherscher Ring, J ein Definitionsideal in A, $Y = \mathrm{Spec}(A)$, Y' die durch J definierte abgeschlossene Teilmenge von Y, $f : X \to Y$ ein separierter Morphismus endlichen Typs und $X' = \bar{f}^1(Y')$. Bezeichnen $\widehat{X}$ bzw. $\widehat{Y}$ und $\widehat{f} : \widehat{X} \to \widehat{Y}$ die Komplettierungen von X bzw. Y längs X' bzw. Y' und von f, so definiert der Funktor $F \rightsquigarrow \widehat{F}$ eine Äquivalenz zwischen der Kategorie der kohärenten O_X-Moduln mit eigentlichem Träger auf $\mathrm{Spec}(A)$ und der Kategorie der kohärenten $O_{\widehat{X}}$-Moduln mit eigenlichem Träger auf $\mathrm{Spf}(A)$.

Ist f eigentlich, so gilt der obige Satz ohne die beiden Einschränkungen "mit eigentlichem Träger".

Hans-Joachim Nastold

Author profile:

Grothendieck, Alexander

Spellings: Grothendieck, A. [74] Grothendieck, Alexander [26] Grothendieck, Alexandre [16] Grothendieck, A. [1]
Author-Id: grothendieck.alexander
Publications: 117 including 32 Book(s) and 41 Journal Article(s)

MSC 2010

55	**14** · Algebraic geometry
15	**00** · General mathematics
13	**18** · Category theory, homological algebra
9	**46** · Functional analysis
4	**01** · History; biography

more ...

Journals

6	Publications Mathématiques
4	Comptes Rendus de l'Académie des Sciences, Paris
3	Bulletin de la Société Mathématique de France
3	Canadian Journal of Mathematics
2	American Journal of Mathematics

more ...

Co-Authors

14	Demazure, Michel
5	Dieudonné, Jean Alexandre
3	Verdier, Jean-Louis
1	Berthelot, Pierre
1	Illusie, Luc

more ...

Publication Years

Zbl 108.09301
Hörmander, Lars
Linear partial differential operators.
Die Grundlehren der mathematischen Wissenschaften. 116.
Berlin- Göttingen-Heidelberg: Springer-Verlag. VII, 285 p. with 1 fig.
(1963).

Das Erscheinen dieser Publikation ist das größte Ereignis auf dem Gebiete der Literatur der partiellen Differentialgleichungen seit einem Vierteljahrhundert, d. h. seit 1937, als der II. Band von Courant–Hilbert erschien. [*R. Courant, D. Hilbert*, Methoden der mathematischen Physik. Bd. 2. (Die Grundlehren d. math. Wiss. in Einzeldarstell. 48) Berlin: Julius Springer. XVI, 549 S., 57 Abb. (1937; Zbl 17.39702; JFM 63.0449.05)].

Das Buch besteht aus drei Teilen = 10 Kapiteln. Der 1. Teil ("Funktionalanalysis") enthält im 1. Kapitel eine elementare, aber elegante Einführung in die Distributionentheorie von L. Schwartz, die gipfelt in den Sätzen vom Paley–Wiener Typus über die Fourier–Laplacesche Transformation der Distributionen, deren singulärer Träger: sing sup $u \subset R(0, r)$. Dabei ist sing sup u eine solche kleinste abgeschloßene Menge K, daß $u \in C^\infty(CK)$. Es werden auch Distributionen auf Mannigfaltigkeiten eingeführt. Im 2. Kapitel werden viele – für die späteren Anwendungen wichtige – Distributionenräume eingeführt und untersucht.

Die eigentliche Theorie der Differentialoperatoren umfaßt Kapitel III–X. Auf diesen knapp 200 Seiten sind ungeheuer viel Fakten und Methoden zusammengepreßt, wobei hervorzuheben ist, daß der Verf. alles streng beweist (es wird nie wegen "schwieriger Beweise" auf die Literatur verwiesen). Schon die Einteilung der Theorie ist für den Nichtspezialisten revolutionär: In allen klassischen Textbüchern wird meistens nur die Theorie der Gleichungen zweiter Ordnung behandelt, wobei man zuerst die (algebraische) Einteilung in Typen (elliptisch, hyperbolisch usw.) vornimmt und dann jeden Typus in einem besonderen Teil behandelt. Verf. dagegen behandelt gleich Operatoren beliebiger Ordnung. Seine Einteilung: konstante Koeffizienten Kapitel III–V; variable Koeffizienten: Kap. VI–X. Selbstverständlich kann man aus dieser Fülle hier nur probeweise einiges hervorheben. Kapitel III: Eine Distribution $E \in \mathcal{D}'(R_n)$ heißt Fundamentallösung des Operators $P(D)$, falls $P(D)\mathcal{E} = \delta_0$ (Diracsches Maß). Es wird die Existenz einer Fundamentallösung (für jeden Operator mit konstanten Koeffizienten) bewiesen. Weiter wird eine Konstruktion der Fundamentallösung mit benötigten lokalen Eigenschaften gegeben. Diese Konstruktion erlaubt dem Verf., die lokale Regularität der Lösungen der Gleichung $P(D)u = f$ zu untersuchen. Ein Satz vom Runge-Typus wird bewiesen und die Lösbarkeit der Gleichung $P(D)u = f$ für $f \in \mathcal{D}'_F(\Omega)$ – Distribution endlicher Ordnung – entwickelt. Kapitel IV: Regularität der Lösungen im Innern des Gebietes. Es werden die bekannten (Hörmanderschen) Bedingungen für die Hypoelliptizität (H. e.) von $P(D)$ aufgestellt: es sind Bedingungen dafür, daß die Gleichung $P(D)u = f$ für $f \in C^\infty$ nur C^∞-Lösungen besitzt. Formuliert in der Sprache der Träger: P^* ist h. e., falls

sing sup u = sing sup $P^*(D)u$, $u \in \mathcal{D}'(\Omega)$. Kapitel V: Cauchy-Problem (konstante Koeffizienten). Es werden u. a. in moderner Form klassische Sätze von Cauchy–Kovalevskaja sowie von Holmgren hergeleitet. Weiter wird ein globaler Satz von F. John bewiesen und eine Charakterisierung der Gleichungen gegeben, für welche das nichtanalytische Cauchy-Problem für beliebige rechte Seite und Cauchyanfangsdaten der Klasse C^∞ lösbar ist.
Kapitel VI: Gleichungen ohne Lösungen. *H. Lewy* [Ann. Math. (2) 66, 155–158 (1957; Zbl 78.08104), erratum. ibid. 68, 202 (1958)] hat entdeckt, daß die Gleichung $\partial u/\partial x_1 + \partial u/\partial x_2 + 2i(x_1 + ix_2)\partial u/\partial x_3 = f$ in jedem offenem $\Omega \subset \mathbb{R}^3$ bei "passender" Wahl von $f \in C^\infty(\mathbb{R}^3)$ keine Lösung besitzt. Verf. verallgemeinert dieses Ergebnis folgendermaßen: Satz. Es sei P von einer Ordnung $\leq m$, woraus folgt, daß der Kommutator $C := P^*P - PP^*$ von einer Ordnung $\leq 2m-1$ ist. Angenommen, für ein $x \in \Omega$ und einen kovarianten Vektor ξ gilt (*) $P_m(x,\xi) = 0$, $C_{2m-1}(x,\xi) \neq 0$, dann gibt es ein $f \in C^\infty(\Omega)$, so daß die Gleichung $Pu = f$ in keiner Umgebung von x eine Lösung – sogar im Sinne der Distributionentheorie – besitzt. Wenn für jedes x in einer dichten Untermenge von Ω (*) für ein ξ erfüllt ist, dann gibt es ein $f \in C^\infty(\Omega)$, so daß die Gleichung $Pu = f$ in keiner offenen Untermenge von Ω eine Distributionenlösung besitzt. Kapitel VII: Differentialoperatoren von konstanter Stärke (k. S.). Es sei

$$\tilde{P}(x,\xi) := \sum_{|\alpha| \geq 0} |P^\alpha(x,\xi)|^2,$$

wobei $P^\alpha(\eta) := i^{|\alpha|}D^\alpha P(\eta)$. Ein Operator $P(x,D)$ ist von k. S. in Ω, falls für $x, y \in \Omega$ gilt $\tilde{P}(x,\xi)/\tilde{P}(y,\xi) \leq c(x,y)$ mit $\xi \in \mathbb{R}^n$. Verf. beweist die Existenz eines Greenschen Operators (eine Verallgemeinerung der klassischen Greenschen Funktion) für ein abstraktes Randwertproblem für einen Operator P von k. S., nämlich: Wenn $P(x,D)$ C^∞-Koeffizienten besitzt und von k. S. in einer Umgebung von x_0 ist, und falls Ω eine hinreichend kleine Umgebung von x_0 ist, dann gibt es eine lineare Abbildung $E : \mathcal{E}'(R_n) \to \mathcal{E}'(R_n)$ mit folgenden Eigenschaften: $P(x,D)Ef = f$ in Ω für $f \in \mathcal{E}(R_n)$; $EP(x,D)u = u$ in Ω für $u \in \mathcal{E}'(\Omega)$.
Kapitel VIII (Differentialoperatoren mit einfachen Charakteristiken) enthält die vielleicht schönsten Ergebnisse des Verf.; wir heben folgende hervor: Zuerst werden Abschätzungen vom Carlemanschen Typus diskutiert, d. h. für eine fixierte Funktion φ:

$$\tau \sum_{|\alpha| \leq m-1} \int |D^\alpha u|^2 \exp(2\tau\varphi)dx \leq$$

$$K_1 \int |P(x,D)u||^2 \exp(2\tau\varphi)dx + K_2 \sum_{|\alpha| \leq m-2} \tau^{2(m-|\alpha|)-1} \int |D^\alpha u|^2 \exp(2\tau\varphi)dx, \qquad (*)$$

$u \in C_0^\infty(\Omega)$, $\tau > \tau_0$. Es sei $N = \text{grad}\varphi(x)$ mit $x \in \Omega$; $\zeta = \xi + i\sigma N$ mit $\xi \in R_n$ und $0 \neq \sigma \in R_1$ genüge der charakteristischen Gleichung $P_m(x,\zeta) = 0$. Mit (*) gilt dann auch die Ungleichung

$$|\zeta|^{2(m-1)} - K_2\sigma^2(|\zeta|^2 + \sigma^2)^{m-2} \le 2K_1\varphi_P''(x,\xi,\sigma),$$

wenn die linke Seite positiv ist. Dabei wurde die folgende Abkürzung benutzt

$$\varphi_P''(x,\xi,\sigma) = \sum_{i,k=1}^{n} \frac{\partial^2\varphi}{\partial x_i \partial x_k} P_m^{(j)}(x,\zeta) \overline{P_m^{(k)}(x,\zeta)} +$$

$$\sigma^{-1}\mathrm{Im}\left(\sum_{1}^{n} P_{m,k}(x,\zeta)\overline{P_m^{(k)}(x,\zeta)}\right),$$

(**)

wo $P_m^{(j)}(x,\xi) = \partial P_m(x,\xi)/\partial\xi_j$, $P_{m,j}(x,\xi) = \partial P_m(x,\xi)/\partial x_j$.

Definition. P heißt prinzipiell normal (principally normal), wenn es einen Differentialoperator Q der Ordnung $m-1$ gibt, so daß

$$C_{2m-1}(x,\xi) = 2\mathrm{Re}P_m(x,\xi)\overline{Q_{m-1}(x,\xi)},$$

$x \in R_n$. Es gilt eine (Teil-)Umkehrung des vorstehenden Satzes: Es sei P prinzipiell normal und es sei $\varphi_P''(x,\xi,\sigma) > 0$, wenn $x \in \Omega$, $0 \ne \xi \in R_n$ und $P_n(x,\xi) = 0$. Ist K eine kompakte Untermenge von Ω, dann gilt für eine Konstante C

$$\sum_{1}^{m-1} \tau^{2(m-j)-1} \int |D^j u|^2 \exp(2\tau\varphi)dx \le$$

$$C \int (|Pu|^2 + \tau^{2m-1}|u|^2 \exp(2\tau\varphi)dx, \qquad u \in C_0^\infty(K).$$

Wenn außerdem $\varphi_P''(x,\xi,\sigma) > 0$ für alle $x \in \Omega$, $\xi \in R_n$, $\sigma \in R$ mit $\sigma \ne 0$ der Gleichung $P_m(x,\xi + i\sigma N) = 0$ genügt, dann gilt

$$\sum_{1}^{m-1} \tau^{2(m-j)-1} \int |D^j u|^2 \exp(2\tau\varphi)dx \le C \int |Pu|^2 \exp(2\tau\varphi)dx, \quad u \in C_0^\infty(K). \quad (***)$$

Bemerkung. $|D^j u|^2 := \sum_{|\alpha|\le j} |D^\alpha u|^2$; bei $P_m(x,\xi + i\sigma N) = 0$ ändert man (**), indem man das zweite Glied ersetzt durch

$$\sigma^{-1}\mathrm{Im} \sum_{1}^{n} P_{m,k}(x,\zeta)\overline{P_m^{(k)}(x,\zeta)} - \mathrm{Re}P_m(x,\zeta)\overline{Q_{m-1}(x,\zeta)}.$$

Aus der Ungleichung (***) bekommt man den folgenden Satz (Eindeutigkeit des Cauchyproblems): Es sei φ eine reelle Funktion in $C^2(\Omega)$ und P ein Differentialoperator, der entweder elliptisch oder prinzipiell normal ist. Weiter sei $x_0 \in \Omega$ und $\mathrm{grad}\varphi(x_0) = N^0 \ne 0$ sowie $\varphi_P''(x_0,\xi,\sigma) > 0$ für $0 \ne \zeta = \xi + i\sigma N^0$, $\xi \in \mathbb{R}_n$, $\sigma \in R_1$, und ζ sei eine Lösung der Gleichung $P_m(x_0,\zeta) = 0$, $\sum_{1}^{n} P_m^{(j)}(x_0,\zeta)N_j^0 = 0$. Dann gibt es eine solche Umgebung Ω'

von x_0, daß jedes $u \in \mathcal{D}'(\Omega)$, das der Gleichung $Pu = 0$ in Ω genügt und auf $\Omega^+ = \{x : x \in \Omega, \varphi(x) > \varphi(x_0)\}$ verschwindet, in Ω_1 identisch verschwindet. Eindeutige Fortsetzung der Singularitäten garantiert der folgende Satz: Es sei φ eine reelle Funktion in $C^2(\Omega)$ und P prinzipiell normal. Weiter sei x_0 ein Punkt in Ω mit $\mathrm{grad}\varphi(x_0) = N^0 \neq 0$, und es sei $\varphi_P''(x_0, \xi, \sigma) > 0$ für $0 \neq \xi \in R_n$ und $P_m(x_0, \zeta) = 0$, $\sum_1^n P_m^{(j)}(x_0, \zeta)N_j^0 = 0$. Dann gibt es eine solche Umgebung Ω' von x_0, daß jedes $u \in \mathcal{D}'(\Omega)$, für welches $Pu \in C^\infty(\Omega)$ und $u \in C^\infty(\Omega^+)$ gilt, in $C^\infty(\Omega)$ enthalten ist.

Im Kapitel IX (Cauchy-Problem für variable Koeffizienten) gelingt es dem Verf. zum großen Teil, Ergebnisse des V. Kapitels auf Operatoren mit variablen Koeffizienten zu übertragen. Kapitel X (Elliptische Randwertaufgaben). In letzter Zeit ist es gelungen, eine richtige Definition des elliptischen Randwertproblems beliebiger Ordnung zu geben. In diesem Kapitel gibt der Verf. eine moderne Fassung dieses Problems: Das Randwertproblem wird betrachtet als eine lineare (stetige) Abbildung eines geeigneten Funktionenraumes in, ein Produkt geeigneter Funktionenräume. Verf. betrachtet gleich kompakte differenzierbare Mannigfaltigkeiten mit Rand. Es werden auch (allgemeine) adjungierte Probleme definiert. Endlichkeit des Indexes der Randwertaufgabe und Stabilität des Indexes bei kleinen Störungen werden bewiesen. Verf. beweist weiterhin die differenzierbare Abhängigkeit der Lösungen elliptischer Randwertaufgaben von den Daten des Problems. Kurz skizziert wird auch eine Übertragung des obigen Resultats auf elliptische Systeme.

Zum Schluß möchte der Ref. einige "kritische" Bemerkungen hinzufügen: Diese Publikation wird mit Sicherheit ein Standardwerk werden, so daß bald eine zweite Auflage nötig sein wird. Dann wäre es wünschenswert, beide Indexe zu vervollständigen: z. B. in dem "Index of notations" fehlen die oft benutzten Bezeichnungen: P^α, $\tilde{P}$, R_n^0, $\mathcal{M}(\Omega; \omega, \mu)$, $\mathcal{M}_0$, $\mathcal{M}_{-\infty}(R_n, R_n; \mu)$. Im "Index": uniqueness of Cauchy problem, continuation of singularities etc. Dem Nichtspezialisten wäre mit kurzen Brückenschlägen zur konventionellen Betrachtungsweise geholfen, z. B. warum der schöne Satz 8.9.1 wirklich Eindeutigkeit des Cauchyproblems ergibt. Doch niemand hat bisher aller Menschen Wünsche befriedigt. ... Wir sind dem Verf. jedenfalls zutiefst dankbar, daß er uns dieses Werk geschenkt hat. *Krzysztof Maurin*

Author profile:

Hörmander, Lars

Spellings: Hörmander, Lars [116] Hörmander, L. [38] Hoermander, L. [2] Hörmander, Lars [1] Hoermander, Lars [1]
Author-Id: hormander.lars
Publications: 158 including 30 Book(s) and 98 Journal Article(s)

MSC 2010

86	**35** · Partial differential equations (PDE)
25	**47** · Operator theory
14	**32** · Functions of several complex variables and analytic spaces
12	**42** · Fourier analysis
12	**58** · Global analysis, analysis on manifolds

more ...

Journals

16	Mathematica Scandinavica
11	Arkiv för Matematik
10	Communications on Pure and Applied Mathematics
9	Acta Mathematica
4	Annals of Mathematics. Second Series

more ...

Co-Authors

3	Melin, Anders
2	Gårding, Lars
2	Sigurdsson, Ragnar
1	Agmon, Shmuel
1	Bernhardsson, Bo

more ...

Publication Years

Zbl 192.04401
Cohen, Paul Joseph
The independence of the continuum hypothesis. I, II.
Proc. Natl. Acad. Sci. USA 50, 1143-1148 (1963); 51, 105-110 (1964).

These two papers contain the solution of the famous Continuum Hypothesis. The author shows that the continuum hypothesis cannot be derived from the axioms of set theory. The proof is based an his method of forcing, which has become since very popular and led to numerous discoveries in the foundations of mathematics.

The author starts with an assumption that there is a countable transitive model M of Zermelo-Fraenkel set theory ZF which satisfies the axiom of constructibility and constructs an extension N of M which is a model of ZF + Axiom of Choice+$2^{\aleph_0} > \aleph_1$. The idea is to obtain N by adjoining to M a sequence $\{a_\delta : \delta < \aleph_\tau\}$ of "generic" sets of integers, where $\aleph_\tau$ is a cardinal number in M, greater than $\aleph_1$.

The key device is the notion of forcing. The forcing language consists of names of all elements of M and of generic sets a_δ, and of expressions using logical symbols and set-theoretical operations. A condition is a finite consistent set of expressions $n \in \widehat{a_\delta}$ or $n \notin \widehat{a_\delta}$. A condition P forces $n \in \widehat{a_\delta}$ if $(n \in a_\delta) \in P$. Similarly, we can define the relation "P forces φ" for any condi-

Author profile:

Cohen, Paul Joseph

Spellings: Cohen, Paul J. [12] Cohen, P.J. [7] Cohen, Paul Joseph [3]
Author-Id: cohen.paul-joseph
Publications: 21 including 2 Book(s) and 16 Journal Article(s)

MSC 2010

7	03 · Mathematical logic
1	00 · General mathematics
1	01 · History; biography
1	30 · Functions of a complex variable
1	43 · Abstract harmonic analysis

more ...

Journals

2	American Journal of Mathematics
2	Bulletin of the American Mathematical Society
2	Proceedings of the American Mathematical Society
1	American Mathematical Monthly
1	Communications on Pure and Applied Mathematics

more ...

Co-Authors

1	Lees, Michael

Publication Years

1950	1960	1970	1980	1990	2000	

tion P and any formula φ of the forcing language. This definition is carried out inside M and has the ollowing properties

(a) for each φ, no P forces both φ and $\neg\varphi$;

(b) if P forces φ and $Q \supset P$ then Q forces φ;

(c) for each φ and each P, there is $Q \supset P$ which decides φ (i.e. Q forces φ or Q forces $\neg\varphi$).

Since M is countable, there is a sequence $P_0 \subseteq P_1 \subseteq \ldots P_s \ldots$ (outside M) such that each formula is decided by some P_s. The extension N is then obtained by adjoining to M the sequence $\{a_\delta : \delta < \aleph_\tau\}$ where $a_\delta = \{n : n \in a_\delta$ belongs to some $P_s\}$. The significance of the forcing method in the construction of N is expressed by the following

Lemma: A formula is true in N if and only if it is forced by some P_s. Using this, the author proves that N is a model of ZF + Axiom of Choice and that $\{a_\delta : \delta < \aleph_\tau\}$ is a sequence of distinct sets of integers. The proof is completed when verified that every cardinal number in M is also a cardinal number in N, so that N satisfies $2^{\aleph_0} > \aleph_1$.

Finally, it is shown how the construction described above yields the relative consistency proof of ZF + Axiom of Choice + $2^{\aleph_0} > \aleph_1$. To verify the truth of a statement in N we need the truth of only finitely many axioms in M; and since every finite collection of axioms of ZF has a countable transitive model, every contradiction in ZF + AC + $2^{\aleph_0} > \aleph_1$ leads to a contradiction in ZF.

Tomas Jech

Zbl 127.09002
Cooley, James W.; Tukey, John W.
An algorithm for the machine calculation of complex Fourier series.
Math. Comput. 19, 297-301 (1965).

Author profile:

Cooley, James W.

Spellings: Cooley, James W. [5] Cooley, J.W. [5]
Author-Id: cooley.james-w
Publications: 10 including 8 Journal Article(s)

MSC 2010

5	**65** · Numerical analysis
4	**42** · Fourier analysis
2	**68** · Computer science
1	**11** · Number theory
1	**20** · Group theory and generalizations

more ...

Journals

2	Journal of Sound and Vibration
2	Mathematics of Computation
1	Advances in Applied Mathematics
1	IBM Journal of Research and Development
1	IEEE Transactions on Acoustics, Speech, and Signal Processing

more ...

Co-Authors

3	Lewis, Peter A.W.
3	Welch, Philip D.
2	Tolimieri, Richard
1	Agarwal, Ramesh C.
1	An, Myoung

more ...

Publication Years

1960	1970	1980	1990

The first review of 1965 simply reads "Ein Zeit und Speicherraum sparendes Verfahren zur elektronischen Berechnung komplexer Fourierreihen mit vorgegebenen Koeffizienten.", i.e. "A time- and memory space-saving algorithm for the electronic computation of complex Fourier series with prescribed coefficients." 45 year later, the result of the short note has grown into manifold applications.

The discrete Fourier transform with N degrees of freedom is equivalent to the evaluation of a polynomial

$$f(z) = \sum_{k=0}^{N-1} \alpha_k z^k.$$

at the N points $1, \omega, \omega^2, \ldots, \omega^{N-1}$ where $\omega = e^{2\pi i/N}$ is the unit root of $z^N - 1 = 0$. The straight-forward computation requires N^2 multiplications.

Cooley noticed that the computing effort can be reduced to (a multiple of) $N(r_1 + r_2 + \cdots + r_n)$ if $N = r_1 r_2 \cdots r_n$. In particular, there is a reduction to $N \log N$ if $N = 2^n$. Encouraged by Garwin, who in his own research was in need of a fast means to compute Fourier transforms, Cooley and Tukey provided a computer code of the algorithm that is now the famous Cooley-Tukey algorithm.

Now there is an uncountable number of papers on the algorithm, and it is even impossible to provide an exhaustive list of the bibliographic notes on it. Therefore the present review can contain only a personal choice of hints to the topic, and the same holds for the references. Today it is unbelievable that Garwin had to put some pressure on the authors before he could convince them to publish a paper on the algorithm.

After seeing the paper by Cooley and Tukey it is clear that Gauß knew the procedure already in 1805 and Runge in 1903 [cf. *C.F. Gauß*, Nachlass: "Theoria interpolationis methodo nova tractata", Werke Band III, 265–327 (Königliche Gesellschaft der Wissenschaften, Göttingen, 1866), here p. 325, and *M.T. Heideman, D.H. Johnson, C.S. Burrus*, Arch. Hist. Exact Sci. 34, 265–277 (1985; Zbl 577.01027)]. Gauß calculated a sine series for describing the motion of the asteroid Juno. It is also hard to read Runge's discussion today. There are more predecessors, but in the pre-computer period there was not such a demand of algorithms for large scale problems.

The fast Fourier transform (FFT) is not only important for problems like signal processing which are directly related to the Fourier transform, it is important for any kind of numerical treatment of convolution problems. The fast computation of the product of large numbers and – in connection with this application – of polynomials of a high degree belong to the important applications.

The Cooley-Tukey algorithm reduces the effort from N^2 to $N \log N$. There are fast algorithms also for other numerical procedures that show that the asymptotic complexity may be smaller than expected from the knowledge of straight-forward procedures. Certainly Strassen and Winograd who investigated such complexity problems, were inspired by the FFT. We mention the fast computation of large matrices. Nevertheless, there is a difference. The fast Fourier transform is highly efficient not only for large N, but also for small N. Moreover, it is more robust against rounding errors than a simple summation in the straight-forward evaluation. Specifically, when multiplying large numbers, the computing time is only increased by a factor depending on $\log \log n$ in order to cope with rounding errors.

In the case $N = 2^n$ one can explain the algorithm to students via a repeated use of a simple even-odd reduction. Starting with an N^2 law, the solution of two problems of the size $N/2$ requires $2(N/2)^2 = N^2/2$ operations, and the repetition eventually leads to the $N \log N$ law. Moreover the bit inversion that is found in the appropriate indexing is a good hint to the duality character.

Indeed, defining $f_\ell := f(\omega^\ell)$ we want to calculate

$$f_\ell = \sum_{k=0}^{N-1} \alpha_k \omega^{k\ell}.$$

The first reduction refers to

$$f_{2\ell} = \sum_{k=0}^{N/2-1} \left(\alpha_k \omega^{2k\ell} + \alpha_{k+N/2} \omega^{2(k+N/2)\ell} \right) = \sum_{k=0}^{N/2-1} \beta_k (\omega^2)^{k\ell},$$

$$\text{where } \beta_k = \alpha_k + \alpha_{k+N/2}.$$

$$f_{2\ell+1} = \sum_{k=0}^{N/2-1} \left(\alpha_k \omega^{2k\ell} \omega^k + \alpha_{k+N/2} \omega^{2(k+N/2)\ell} \omega^{k+N/2} \right) = \sum_{k=0}^{N/2-1} \gamma_k (\omega^2)^{k\ell},$$

$$\text{where } \gamma_k = \omega^k (\alpha_k - \alpha_{k+N/2}).$$

Author profile:

Tukey, John W.

Spellings: Tukey, John W. [92] Tukey, J.W. [9] Tukey, J. W. [6] Tukey, John.W. [1] Tukey, John [1]
Author-Id: tukey.john-w
Publications: 109 including 20 Book(s) and 76 Journal Article(s)

MSC 2010

45	**62** · Statistics
5	**01** · History; biography
4	**65** · Numerical analysis
3	**68** · Computer science
2	**00** · General mathematics

more ...

Journals

24	Annals of Mathematical Statistics
6	JASA. Journal of the American Statistical Association
4	Biometrika
4	Technometrics
3	Bulletin of the American Mathematical Society

more ...

Co-Authors

10	Mosteller, Frederick
6	Hoaglin, David C.
5	Morgenthaler, Stephan
2	Basford, Kaye Enid
2	Ciminera, Joseph L.

more ...

Publication Years

Since ω^2 is the unit root of $z^{N/2} - 1 = 0$, the values for even indices and for odd indices are indeed given in terms of a discrete Fourier transform of half the size. The computing effort for the reduction is proportional to N. In the second step the procedure is repeated for two problems of size $N/2$, and after $\log_2 N$ steps with an $O(N)$ effort we are done. The following table shows for $N = 8$ the repeated splitting into subproblems and how the indices of the function values are reordered during the $3 = \log_2 8$ steps. The indices in the third column are written as binary numbers. If we read the bits from the right to the left, we get the numbers $0, 1, 2, \ldots, 7$ in the natural order and see the bit inversion.

$$
\begin{array}{ccc}
0 & 0 & 0\ 000 \\
1 & 2 & 4\ 100 \\
2 & 4 & 2\ 010 \\
3 & 6 & 6\ 110 \\
4 & 1 & 1\ 001 \\
5 & 3 & 5\ 101 \\
6 & 5 & 3\ 011 \\
7 & 7 & 7\ 111
\end{array}
$$

In the case of a decomposition $N = r_1 r_2 \cdots r_n$ with factors $r_i \neq 2$, considerations from group theory yield a better understanding of the mathematics behind the Cooley-Tukey algorithm. From an algebraic point of view, the cyclic discrete Fourier transform can be explained in terms of group algebras of cyclic groups and their representations. The elements $1, z, z^2, \ldots, z^{n-1}$ form a $\mathbb{C}$-basis of the algebra $\mathbb{C}[z]/(z^N - 1)$, and this basis is a cyclic group of order N. This concept can be generalized to fast convolution procedures in arbitrary finite groups. In this context we mention that the braid group was considered in connection with the fast multiplication of large matrices and its complexity. The literature on the algebraic and group theoretic aspects makes only a very small portion compared to the algorithmic aspects and the applications in signal theory.

The Cooley-Tukey algorithm is not only important for the applications in connection with the fast Fourier transform and convolution, it also gave rise to interesting mathematics.

Karl Stumpff (1965); Dietrich Braess (Bochum) (2010)

Zbl 158.40802
Kuratowski, K.
Topology. Vol. I. New edition, revised and augmented.
New York-London: Academic Press; Warszawa: PWN-Polish Scientific Publishers. XX, 560 p. (1966).

Seit dem ersten Erscheinen des Kuratowskischen 2-bändigen Werkes zur allgemeinen Topologie sind inzwischen über 30 Jahre vergangen [Topologie. I. Espaces metrisables, espaces complets. (Monogr. mat. 3) Warszawa, Lwow: Subwencji Funduszu Kultury Narodowej. IX, 285 S. (1933; Zbl 8.13202; JFM 59.0563.02); II: Monografie Matematyczne T. XXI. Warszawa: Seminarium

Matematyczne Uniwersytetu. VIII, 444 p. (1950; Zbl 0041.09604)]. In dieser verflossenen Zeit hat sich das Buch in dem gesamten Schrifttum zur allgemeinen Topologie einen hervorragenden (wenn nicht gar überhaupt den hervorragendsten) Platz gesichert. Mit der jetzt vorliegenden englischen und russischen Fassung wird das Standardwerk, bisher in französischer Sprache, seinen weltweiten Einfluß noch erheblich vergrößern, indem es nun auch der jüngeren Generation bequem zugänglich ist und seine Funktion als Lehr- und Handbuch in breiteren Kreisen ausüben kann. Die Neufassung des ersten Bandes betrifft keineswegs nur die Sprache, vielmehr wird dein modernen Trend in der Entwicklung der topologischen Forschung Rechnung getragen. Das geschieht aber nicht so sehr in der Weise, daß die neueren Dinge stets ausführlich dargestellt werden, sondern durch eine Reihe von Hinweisen skizziert der Verf. den weiteren Verlauf. Durch dieses Verfahren sind die Vorzüge des ursprünglichen Kuratowskischen Buches sämtlich erhalten geblieben, nämlich eine Darstellung zu liefern, die im wesentlichen einen elementaren Charakter hat, die das Eigentliche an den Dingen hervortreten läßt und die angestrebte Allgemeinheit der Natur der Sache anpaßt. Der Gesamtaufbau des ersten Bandes gliedert sich in 3 Hauptabschnitte, die den Grundbegriffen der topologischen Räume, der metrischen Räume und der vollständigen Räume gewidmet sind. Vorausgeht diesen Kapiteln eine Einführung in die elementaren Begriffsbildungen der Mengenlehre und mathematischen Logik, abgeschlossen wird der Textteil des Buches mit einer kurzen Skizze von A. Mostowski über einige Anwendungen der Topologie auf Fragen der mathematischen Logik (z. B. topologische Beweise des Gödelschen Vollständigkeitstheorems und des Löwenheim Skolemschen Modelltheorems) sowie einigen Hinweisen von R. Sikorski auf Anwendungen der Topologie in der Funktionalanalysis.

Eine knappe Aufzählung der im Hauptteil behandelten Gegenstände mag einen ungefähren Eindruck von diesem reichhaltigen Buch verschaffen. Topologische Räume (150 Seiten): Die Abschließungsoperation (Kuratowskische Hüllenaxiome), abgeschlossene und offene Mengen. Begrenzung und Inneres von Mengen. Umgebungen. Dichte und nirgendsdichte Mengen. Häufungspunkte. Magere Mengen. Bairesche Eigenschaft. Stetigkeit und Homöomorphie. Vollständig reguläre und normale Räume. Produkträume. Raum der abgeschlossenen Mengen. Halbstetige Abbildungen und Quotientenräume.

Metrische Räume (216 Seiten): Konvergenzklassen. Allgemeine Begriffsbildungen für metrische Räume (hierbei werden auch die uniformen Räume und Nachbarschaftsräume gestreift). Separable Räume. Mächtigkeitsfragen. Dimensionsproblexnatik, insbesondere Anfänge der Polyedertopologie. Zusammenhänge der metrischen Topologie mit der deskriptiven Mengenlehre und Theorie der reellen Funktionen.

Vollständige Räume (127 Seiten): Grundlegende Begriffe, Cauchyfolgen und Vervollständigung. Bairesches Theorem. Fortsetzung von Funktionen. Zusammenhänge der vollständigen Räume mit dem Raum der Irrationalzahlen. Fortführung der deskriptiven Mengenlehre in Verbindung mit vollständigen Räumen.

Jürgen Flachsmeyer

Author profile:

Kuratowski, Casimir

Spellings: Kuratowski, C. [121] Kuratowski, K. [70] Kuratowski, Casimir [53] Kuratowski, Kazimierz [19] Kuratowski, Kasimierz [1] Kuratowski, C [1]
Author-Id: kuratowski.casimir
Publications: 265 including 31 Book(s) and 217 Journal Article(s)

MSC 2010

37	54 · General topology
18	03 · Mathematical logic
7	28 · Measure and integration
6	01 · History; biography
4	26 · Real functions

more ...

Journals

109	Fundamenta Mathematicae
9	Comptes Rendus de l'Académie des Sciences. Paris
5	Colloquium Mathematicum
3	Bulletin of the American Mathematical Society
2	Annali di Matematica Pura ed Applicata. Serie Quarta

more ...

Co-Authors

6	Knaster, Bronisław
6	Mostowski, Andrzej Włodzimierz
6	Sierpiński, Wacław
3	Banach, Stefan
3	Eilenberg, Samuel

more ...

Publication Years

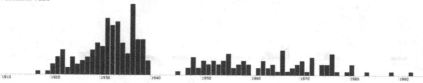

Zbl 146.18802
Puppe, Dieter
Stabile Homotopietheorie. I.
Math. Ann. 169, 243-274 (1967).

Die stabile Homotopietheorie befaßt sich mit der Frage, eine geeignete Kategorie zu finden, in der man die stabilen Phänomene besonders gut studieren kann. Die erste Kategorie dieser Art ist die S-Kategorie von Spanier und J. H. C. Whitehead (hier SWh_0) genannt. Sie hat den Nachteil, daß sie nicht genügend viele Eilenberg-MacLane-Objekte enthält: Zu einer endlich erzeugten Gruppe π gibt es in SWh_0 kein $K(\pi, n)$; ist $\pi = Q = $ rationale Zahlen, so gibt es ein $K(Q, n)$. In dieser Arbeit werden eine Reihe von Kategorien auf ihre Eignung hin untersucht. Zum Beispiel: Sph_0 Objekte = Spektren = Folgen $\{A_i\}$ von CW-Komplexen mit $A_i \wedge S^1 \subseteq A_{i+1}$; Abbildungen = Homotopieklassen von Folgen $\{f_i\}$, $f_i : A_i' \to B_i$ mit gewöhnlicher Verträglichkeitsbedingung bezüglich Einhängung. Sah: Objekte wie oben, Abbildungen auch wie oben, nur sollen jetzt die einzelnen Folgenglieder nur bis auf Homotopie mit der Einhängung verträglich sein. Diese Homotopie ist mit vorgegeben: Sth_0:

108

Objekte = stabile Spektren, d.h. $(A_i \wedge S^1)^{2i+n} = A_{i+1}^{2i+n}$ für ein gewisses n. Abbildungen ähnlich wie oben.

Author profile:

Puppe, Dieter

Spellings: Puppe, Dieter [31] Puppe, D. [1]
Author-Id: puppe.dieter
Publications: 32 including 2 Book(s) and 21 Journal Article(s)

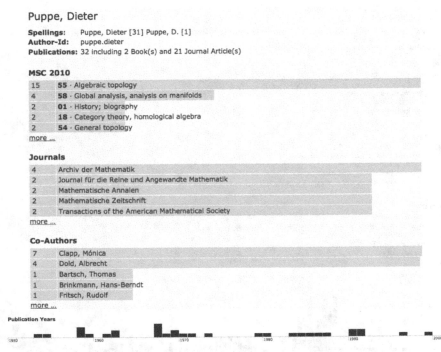

MSC 2010

15	55 · Algebraic topology
4	58 · Global analysis, analysis on manifolds
2	01 · History; biography
2	18 · Category theory, homological algebra
2	54 · General topology

more ...

Journals

4	Archiv der Mathematik
2	Journal für die Reine und Angewandte Mathematik
2	Mathematische Annalen
2	Mathematische Zeitschrift
2	Transactions of the American Mathematical Society

more ...

Co-Authors

7	Clapp, Mónica
4	Dold, Albrecht
1	Bartsch, Thomas
1	Brinkmann, Hans-Berndt
1	Fritsch, Rudolf

more ...

Publication Years

1950   1960   1970   1980   1990   2000

Nachdem einige Ergebnisse über diese Kategorien formuliert worden sind, werden allgemeine "stabile Kategorien" erklärt und nachgewiesen, daß die Kategorie Sth stabil ist. Eine stabile Kategorie $\mathfrak{K}$ ist zunächst $\mathbb{Z}$-graduiert, d.h. $\mathfrak{K}(A, B)$ ist eine Folge von $\mathfrak{K}_p(A, B)$, $p \in \mathbb{Z}$ so daß bei der Komposition von Abbildungen $\mathfrak{K}_p(A, B) \times \mathfrak{K}_p(B, C) \to \mathfrak{K}_{p+q}(A, C)$ sich die Grade addieren. Ferner ist $\mathfrak{K}$ präadditiv, d.h. $\mathfrak{K}_p(A, B)$ ist eine abelsche Gruppe und das Distributivgesetz erfüllt. Entscheidend ist die Existenz einer Klasse C von (sogenannten stabilen) Dreiecken $A \xrightarrow{f} B \xrightarrow{g} C \xrightarrow{h} A$ in $\mathfrak{K}$, wo $|f| + |g| + |h| = -1$ ist $[|f| = p$ wenn $f \in \mathfrak{K}(A, B)]$. Von diesen Dreiecken werden eine Reihe von Eigenschaften verlangt. Verf. untersucht nun zweierlei: 1. Es ist Sth eine stabile Kategorie, wenn man als exakte Dreiecke gerade die nimmt, die durch die Kegelbildung erklärt werden und $\mathrm{Sth}_p(A, B)$ durch $\mathrm{Sth}_0(\Sigma^p A, B)$, $\Sigma^p A_i = A + i + p$ erklärt wird. 2. Über (abstrakt gegebene) stabile Kategorien werden eine Reihe von Sätzen bewiesen: Zum Beispiel 21. $\mathfrak{K}$ ist additiv; 22. Es gibt exakte Folgen (Puppe-Folgen) in $\mathfrak{K}$. 23. Es gibt einen Funktor $\Sigma^n : \mathfrak{K} \to \mathfrak{K}$ $(n \in \mathbb{Z})$ und eine natürliche Äquivalenz zwischen Σ^n und der Identität vom Grade n.

Unten den weiteren Ergebnissen, die in dem vorliegenden ersten Teil der Arbeit bewiesen werden, verdienen erwähnt zu werden: 3. Die Kategorie SWh$_0$ kann in Sth$_0$ eingebettet werden, wenn man sich auf endlichdimensionale CW-Komplexe bei der Definition von SWh$_0$ beschränkt; 4. Der "Aushängungssatz" für Sth.

In einem letzten Abschnitt werden Anwendungen auf Kettenkomplexe gegeben. Ein zweiter Teil wird angekündigt. Einige der hier bewiesenen Sätze sind in dieser oder jener Form, meist unpräzise, bekannt, finden sich aber nirgends in der Literatur (z.B. der Satz über die Nichtexistenz von Eilenberg- und MacLane-Objekten in SWh$_0$). Als unmittelbarsten Vorläufer hat die Arbeit das Buch von *J. F. Adams*, Stable homotopy theory (Berlin-Heidelberg-New York 1966). Von Interesse wäre der Zusammenhang mit den anderen kürzlich bekannt gewordenen Untersuchungen in dieser Richtung, z.B. mit Boardman, mit der Theorie der simplizialen Spektren im Sinne von Kan und G. Whitehead sowie zu den Arbeiten von Burghelea und Deleanu.   *Friedrich-Wilhelm Bauer*

Zbl 164.24001
Atiyah, Michael F.; Singer, Isadore Manuel
The index of elliptic operators. I.
Ann. Math. (2) 87, 484-530 (1968); Russian translation in Usp. Mat. Nauk 23, No.5 (143), 99-142 (1968).

The earlier proof of the Atiyah-Singer index theorem as given in the book by *R. S. Palais* [Seminar on the Atiyah-Singer index theorem. With contributions by M. F. Atiyah, A. Borel, E. E. Floyd, R. T. Seeley, W. Shih and R. Solovay. Princeton, N. J.: Princeton University Press (1965, Zbl 137.17002)] used cobordism theory and was in this respect modelled on the proof of the Riemann-Roch theorem due to the reviewer [New topological methods in algebraic geometry. Berlin-Göttingen-Heidelberg: Springer-Verlag (1956; Zbl 0070.16302)]. This proof did not lend itself to certain generalizations (for example in the equivariant case) because the corresponding cobordism theories are not known.

The index theorem of the paper under review includes the equivariant case: Suppose a compact Lie group G operates differentiably on a compact differentiable manifold X such that the action is compatible with a linear elliptic problem an X. Then the index of this elliptic problem is an element of the representation ring $R(G)$. In the simplest case the elliptic problem is given by an elliptic linear differential operator $D : C^\infty(E) \to C^\infty(F)$ where E, F are complex vector bundles over X. The group G operates on the kernel and on the cokernel of D. Thus we have two finite-dimensional representations of G. The difference of the two representations as elements of $R(G)$ is the index of the elliptic problem. The symbol of such an elliptic problem is an element of $K(BX, SX)$ if one forgets about the G-action. (For the notation see the review of the book of Palais, loc. cit.) If a G-action is given, then equivariant K-theory must be used and the symbol is an element of $K_G(BX, SX)$. The authors define

110

Author profile:

Atiyah, Michael Francis

Spellings: Atiyah, Michael F. [120] Atiyah, Michael [88] Atiyah, M.F. [13] Atiyah, M. [5] Atiyah, Michael Francis [2]
Author-Id: atiyah.michael-francis
Publications: 228 including 31 Book(s) and 131 Journal Article(s)

MSC 2010

64	**58** · Global analysis, analysis on manifolds
54	**53** · Differential geometry
52	**81** · Quantum Theory
39	**01** · History; biography
35	**57** · Manifolds and cell complexes

more ...

Journals

9	Annals of Mathematics. Second Series
9	Topology
8	Bulletin of the London Mathematical Society
5	Inventiones Mathematicae
5	The Quarterly Journal of Mathematics. Oxford Second Series

more ...

Co-Authors

18	Bott, Raoul H.
17	Singer, Isadore Manuel
10	Hirzebruch, Friedrich Ernst Peter
8	Hitchin, Nigel J.
7	Patodi, V.K.

more ...

Publication Years

Author profile:

Singer, Isadore Manuel

Spellings: Singer, I.M. [85] Singer, Isadore M. [6] Singer, Isadore [6] Singer, I. M. [3] Singer, Isadore Manuel [1]
Author-Id: singer.isadore-manuel
Publications: 101 including 18 Book(s) and 65 Journal Article(s)

MSC 2010

43	**58** · Global analysis, analysis on manifolds
28	**53** · Differential geometry
18	**81** · Quantum Theory
15	**00** · General mathematics
13	**57** · Manifolds and cell complexes

more ...

Journals

7	Annals of Mathematics. Second Series
7	Communications in Mathematical Physics
6	Journal of Differential Geometry
4	American Journal of Mathematics
4	Proceedings of the National Academy of Sciences of the United States of America

more ...

Co-Authors

17	Atiyah, Michael Francis
6	Jerison, David S.
5	Baulieu, Laurent
4	Bott, Raoul H.
4	Hoffman, Kenneth

more ...

Publication Years

K and K_6 for a locally compact space as the corresponding reduced groups of the one-point-cpactification of the space. With this understanding we have $K_G(BX, SX) = K_G(TX)$ where TX is the total space of the covariant tangent bundle. Using "enough operators" (pseudo-differential operators) and homotopy properties which ensure that the index depends only on the symbol the authors define the analytical index a-ind : $K_G(TX) \to R(G)$. This is a homomorphism of $R(G)$-modules and has by its very definition the property that the index of an elliptic problem equals a-ind of its symbol. The definition of an $R(G)$-homomorphism t-ind : $K_G(TX) \to R(G)$ in topological terms is given in §3. The "index theorem" is the main theorem 6.7 of the paper and asserts that a-ind and the topological index t-ind coincide. For the definition of t-ind and for the investigation of properties of a-ind and t-ind the following construction is fundamental. Suppose X, Y are G-manifolds with $X \subset Y$ and X compact. Then $TX \subset TY$ and the normal bundle W of $TX \subset TY$ equals $N \oplus N$ lifted to TX where N is the normal bundle of X in Y. Since $N \oplus N$ carries a complex structure, so also does W. The de Rham complex of the exterior powers of W can be tensoredith any complex of vector bundles over TX lifted to W which has compact support in TX. The result is a complex of veetor bundles over W with compact support, because the de Rham complex is canonically trivialized outside a zero-section of W. In this way we get a homomorphism $K_G(TX) \to K_G(W)$. Since W may be regarded as a tubular neighborhood of TX in TY and since the one-point-compactification of TY maps onto the one-point-compactification of W, we have a homomorphism $K_G(W) \to K_G(TY)$. The composition is an $R(G)$-homomorphism $i_! : K_G(TX) \to K_G(TY)$ where $i : X \to Y$ denotes the embedding.
(Of course, many details are omitted in this review. Constructions using the alternating sum of the exterior powers occur in several earlier papers of Atiyah and others.) To define t-ind for a compact differentiable G-manifold X we embed X in a real representation space E of G. This is always possible [R. S. Palais, J. Math. Mech. 6, 73–678 (1957; Zbl 86.02603)]. Let i be the embedding. We have $i_! : K_G(TX) \to K_G(TE)$. Under the embedding j of the origin P in E we have $j_! : K_G(TP) = K_G(P) = R(G) \to K_G(TE)$ $j_!$ is an isomorphism. This is a special rase of the equivariant form of the Bott periodicity theorem [M. F. Atiyah and it D. W. Anderson, K-theory. With reprints of M. F. Atiyah: Power operations in K-theory. New York-Amsterdam: W. A. Benjamin, Inc. (1967; Zbl 159.53302); M. F. Atiyah, Q. J. Math., Oxf. II. Ser. 19, 113–140 (1968; Zbl 159.53501)]. t-ind is defined by $j_! \cdot (t\text{-ind}) = i_!$. The authors show that the definiton is independent of the choice of the embedding. t-ind is the identity of $R(G)$ if X is a point (A1) and the diagram

$$K_G(TX) \xrightarrow{\;i_!\;} K_G(T_Y)$$
$$\searrow_{t\text{-ind}} \qquad \swarrow_{t\text{-ind}} \qquad\qquad \text{(A2)}$$
$$R(G)$$

commutes for any inclusion $i : X \rightarrow Y$ with X, Y compact G-manifolds. An index function ind is given if we have for every compact differentiable manifold X an $R(G)$-homomorphism $K_G(TX) \overset{\text{ind}}{\rightarrow} (G)$. If such an index function ind satisfies (A1) and (A2) then ind = t-ind (proposition 4.1). For the analytical index (A1) is trivial. To prove the main theorem, axiom (A2) has to be checked for the analytical index. This is not easy because for an operator D an X with symbol $\gamma(D)$ we have to construct an operator on Y with symbol $i_! \gamma(D)$ and show that the index of this newperator equals the index of D. This cohstruction is the essential analytical part of the proof of the main theorem. "Once this has been done, we can take Y to be a sphere, and the general index theorem is reduced to the case of operators on the sphere. For these the problem is easily solved." At this point one recognizes that the whole proof has "in spirit, at least" much in common with Grothendieck's proof of the Riemann-Roch theorem. Since it is difficult to verify (A2) directly it is shown that certain axioms (B1), (B2") and (B3) imply (A2) for any index function. In §8 the axioms (B1) and (B2") are proved for the analytical index; in §9 the axiom (B3) is proved. A special case of (B3) is the behaviour of the index function if one takes the cartesian product of two G-manifolds with elliptic problems. The analytical index behaves multiplicatively in this case. (B3) generalizes this multiplicative property to differentiable fibre bundles. In this case the bundle of "indices along the fibres" may not be trivial over the bare and enters essentially in the formulation. (B1) is an excision axiom: Let U be a (non-compact) G-manifold and $j : U \rightarrow X$, $j' : U \rightarrow X'$ two open G-embeddings into compact G-manifolds X, X'. Then the following diagram commutes.

$$K_G(TX)$$

$$K_G(TU) \qquad \qquad R(G)$$

$$K_G(TX')$$

with j^*, ind (upper), j'^*, ind (lower).

Observe that the one-point-compactifications of TX and TX' map onto the one-point conpactification of TU. By these maps j^* and j'^* are induced. The excision property of the analytical index was observed by R. T. Seeley [Trans. Am. Math. Soc. 117, 167–204 (1965; Zbl 135.37102)]. The axiom (B2") is a normalisation axiom for certain operators on S^1 and S^2. Information on operators on other spheres follows by using excision and the multiplicative property. The idea of proof for the essential property (A2) of the analytical index is sketched by the authors as follows (§1): let $i : X \rightarrow Y$ be an inclusion of compact manifolds (we forget the G-action). Let U be a tubular neighborhood of X and Z its double. Then the excision property (B1) shows that ind $i_! A$ = ind $k_! A$ where $A \in (TX)$ and $k : X \rightarrow Z$ is the inclusion in the double. Z is fibred over X by spheres. The multiplicative property and the information on spheres gives the desired result ind $i_! A$ = ind A. This paper contains an impressive amount of analysis. The theory of pseudo-differential operators (Hörmander, Kohn-Nirenberg, Seeley) is essential to have "enough operators" to realize all elements of $K_G(TX)$ as symbols and to carry through all constructions. The analytical index was "calculated" in this paper by topological terms (topological index defined by K-theory). In the following papers II and III this topological index will be interpreted in two steps. In II the topological index is expressed in terms of fixed-point sets of G. This is done in K-theory. It leads to a general "Lefschetz fixed-point theorem" where the fixed-point set of an element of G is a disjoint union of submanifolds. For isolated fixed points this is contained in the fixed point theorem of *M. F. Atiyah* and *R. Bott* [Ann. Math. (2) 86, 374–407 (1967; Zbl 161.43201)], a theorem on differentiable maps g with isolated simple fixed points, where g need not be invertible. In III the result is reformulated in cohomological terms. If G is the identity, this gives the index theorem in its well-known cohomological form (Palais, loc. cit.). In general, it gives the cohomological form of the fixed-point theorem. *Friedrich Hirzebruch (Bonn)*

Zbl 191.17903
Knuth, D.E.
The art of computer programming. Vol. 1: Fundamental algorithms.
Addison-Wesley Series in Computer Science and Information Processing.
London: Addison-Wesley Publishing Company. XXII, 634 p. (1968).

Author profile:

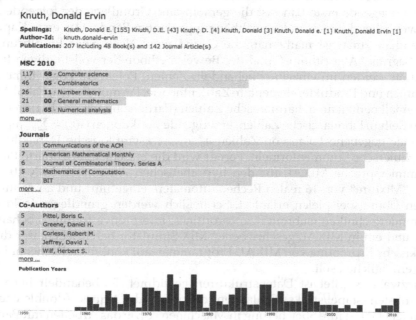

Knuth, Donald Ervin

Spellings: Knuth, Donald E. [155] Knuth, D.E. [43] Knuth, D. [4] Knuth, Donald [3] Knuth, Donald e. [1] Knuth, Donald Ervin [1]
Author-Id: knuth.donald-ervin
Publications: 207 including 48 Book(s) and 142 Journal Article(s)

MSC 2010

117	68 · Computer science
46	05 · Combinatorics
26	11 · Number theory
21	00 · General mathematics
18	65 · Numerical analysis

more ...

Journals

10	Communications of the ACM
7	American Mathematical Monthly
6	Journal of Combinatorial Theory. Series A
5	Mathematics of Computation
4	BIT

more ...

Co-Authors

5	Pittel, Boris G.
4	Greene, Daniel H.
3	Corless, Robert M.
3	Jeffrey, David J.
3	Wilf, Herbert S.

more ...

Publication Years

Dies ist der erste Band eines geplanten 7-bändigen Werkes über die "Kunst des Programmierens": I Fundamental Algorithms: 1. Basic concepts, 2. Information Structures, II Seminumerical Algorithms: 3. Random Numbers, 4. Arithmetic, III Sorting and Searching: 5.Sorting Techniques, 6.Searching Techniques, IV Combinatorial Algorithms: 7. Combinatorial Searching, 8. Recursion, V Syntactic Algorithms: 9. Lexical Scanning, 10. Parsing Techniques, VI Theory of Languages: 11. Mathematical Linguistics, VII Compilers: 12. Programming Language Translation.

Es ist gedacht sowohl als Nachschlagewerk als auch als Lehrbuch für Vorlesungen oder zum Selbststudium. Deshalb sind in den Text eine Vielzahl von Übungen, meist mit Lösungen, eingefügt. Die Monographie richtet sich keineswegs nur an Programmierungsexperten. Es ist im Gegenteil das erklärte Ziel des Verfassers, Programmierungstechniken denen zugänglich zu machen, die auf anderen Gebieten arbeiten und mit Erfolg Digitalrechner für ihre Probleme benutzen könnten. Gewisse Vorkenntnisse werden jedoch benötigt. Nach Meinung des Verfassers ist als Leser geeignet, wer "mindestens vier Programme für mindestens einen Digitalrechner geschrieben und getestet hat". Den Gegenstand des Buches bezeichnet der Verfasser als "Analysis der Algorithmen", worunter er die Theorie der Eigenschaften von speziellen Computeralgorithmen versteht. Dabei liegt sein Hauptaugenmerk auf der sogenannten nichtnumerischen Analysis, also der Entwicklung von "software" für Automaten, von Methoden für das Sortieren, für Sprachübersetzungen, für die Lösung abstrakter mathematischer Probleme und für das Beweisen mathematischer Sätze mit Hilfe von Rechenmaschinen. Der vorliegende erste Band ist die gemeinsame Grundlage der folgenden, als weitgehend voneinander unabhängig geplanten Bände. Das erste Kapitel enthält zunächst mathematische Grundlagen: Nach einer Einführung des Begriffs "Algorithmus" und der Beweismethode der vollständigen Induktion werden im einzelnen behandelt: Zahlen, Potenzen, Logarithmen, Summen und Produkte, elementare Zahlentheorie, Permutationen, Fakultät, Binomialkoeffizienten, harmonische Zahlen (Partialsummen der harmonischen Reihe), Fibonaccische Zahlen, erzeugende Funktionen $G(z) = \sum_{n=0}^{\infty} a_n z^n$ für eine gegebene Folge von Zahlen $a_0, a_1, \ldots$, asymptotische Darstellungen, Eulersche Summenformeln. Ferner wird die maschinenorientierte Programmiersprache MIXAL für den hypothetischen Automaten MIX 1009, eine "Mixture" von 16 realen Rechenautomaten, eingeführt und an zahlreichen Übungsbeispielen erläutert. Schließlich werden grundlegende Programmiertechniken (subroutines, coroutines, interpretive routines) behandelt und ein in MIXAL geschriebener MIX-Simulator beschrieben, der die praktische Erprobung von MIXAL-Programmen auf konkreten Rechenanlagen ermöglichen soll.

Das zweite Kapitel ist Datenstrukturen gewidmet. Es behandelt lineare Datenlisten (stapel/stacks), Schlangen (queues) und "Deques" (double-ended-queues) und ihre Speicherung in Maschinen sowie das Arbeiten mit Bäumen. Welche Bedeutung den Übungsaufgaben in diesem Buch beigemessen

wird, erkennt man daraus, dass die Lösungen (mit Angabe des Lösungs-
weges bzw. Beweisganges) 140 Seiten einnehmen. Der Anhang enthält ein
Verzeichnis der verwendeten Bezeichnungen sowie Tabellen von wichtigen Konstanten (dezimal und oktal) und der harmonischen, Bernoullischen
und Fibonaccischen Zahlen. Hervorzuheben ist schließlich der ausführliche
Index (17 Seiten), der die Eignung des Buches als Nachschlagewerk unter-
streicht.   *Gerhard Maeß*

Zbl 191.18001
Knuth, Donald E.
The art of computer programming. Vol. 2: Seminumerical algorithms.
Addison-Wesley Series in Computer Science and Information Processing.
Reading, Mass.: Addison-Wesley Publishing Company. XII, 624 p. (1969).

Der vorliegende zweite Band enthält die Kapitel 3 (Random Numbers) und
4 (Arithmetic) der geplanten 7-bändigen Monographie über die "Kunst des
Programmierens" [The art of computer programming. Vol. 1: Fundamental algorithms. Addison-Wesley Series in Computer Science and Information Processing. London: Addison-Wesley Publishing Company (1968; Zbl
191.17903)]. Nach einem Überblick über die Verwendung von Zufallszahlen
zur Simulation realer Vorgänge, zur Stichprobenauswahl oder für Monte-
Carlo Methoden der Numerischen Mathematik wird auf die Schwierigkeiten
hingewiesen, zufällige Zahlenfolgen durch determinierte Rechenschritte
aufzustellen. Es werden verschiedene Methoden der Erzeugung vorn Zufallszahlen und statistische Tests zum Nachweis des zufälligen Charakters
behandelt und Transformationen gleichverteilter Zufallszahlen in andere
Typen von Zufallsgrößen (zufällige Wahl unter k Möglichkeiten, Bestimmung eines Zeitpunkts für den Eintritt eines zufälligen Ereignisses, stetige
Verteilungen) angegeben.
Kapitel 4 behandelt verschiedene Systeme der Zahlendarstellung und den
Übergang von einem System zum anderen, Probleme der Gleitkommarech-
nung, das Rechnen mit ganzen Zahlen und Brüchen, die Zerlegung in Prim-
faktoren und die Bestimmung des größten gemeinsamen Teilers. Ferner ent-
hält es zwei Abschnitte über Polyname (Division, Zerlegung in Faktoren,
Berechnung von Potenzen, Berechnung von Polynomen: Hornerschema)
und Potenzreihen (Summe, Produkt, Quotient).
Wie bereits im ersten Band enthält jeder Abschnitt eine Vielzahl von
Übungsaufgaben, deren ausführliche Lösungen im Anhang angegeben sind.
Um das Buch weitgehend unabhängig vom ersten Band zu machen, wird
am Schluss noch einmal auf den hypothetischen Rechner MIX mit seiner bei
vielen Beispielen verwendeten Programmiersprache MIXAL eingegangen.
Schließlich enthält auch der 2. Band Tabellen wichtiger Konstanten (dezimal
und oktal), die Erklärung der verwendeten Bezeichnungen und einen aus-
führlichen Index.   *Gerhard Maeß*

Zbl 302.68010
Knuth, Donald E.
The art of computer programming. Vol. 3: Sorting and searching.
Addison-Wesley Series in Computer Science and Information Processing. Reading, Mass. etc.: Addison-Wesley Publishing Company. XI, 722 p. (1974).

Der 3. Band [vgl. Bd. 1, Fundamental algorithms (1968; Zbl 191.17903) und Bd. 2, Seminumerical algorithms (1969; Zbl 191.18001)] enthält die Kapitel 5 (Sorting) und 6 (Searching) der geplanten 7-bändigen Monographie über die "Kunst des Programmierens". Die stürmische Entwicklung gerade dieser Teilgebiete der Datenverarbeitung in den letzten Jahren verzögerte zwar das Erscheinen des Buches um fast 3 Jahre, führte aber zu einer erstaunlichen Vielfalt des aufgenommenen Stoffes. Der Verf. umreißt die behandelten Probleme durch die folgenden Fragen: Wie findet man gute Algorithmen? Wie können gegebene Algorithmen und Programme verbessert werden? Wie kann die Effektivität von Algorithmen mathematisch analysiert werden? Wie kann man rationall zwischen verschiedenen Algorithmen für das gleiche Problem wählen? In welchem Sinne können Algorithmen optimal sein? Wie können externe Speicher effektiv eingesetzt werden?
Inhaltlich schließt sich der vorliegende Band an Kapitel 2 (Information Structures) des 1. Bandes an. Kap. 5 (Sorting) enthält neben einem Abschnitt über Permutationen und einen über optimales Sortieren die beiden Hauptabschnitte "Internal Sorting" und "External Sorting". Kap. 6 (Searching) ist in folgende Abschnitte untergliedert: Sequential Searching, Searching by Comparison of Keys, Digital Searching, Hashing, Retrieval an Secondary Keys.
Das Buch eignet sich sowohl als Grundlage für Vorlesungen unterschiedlichen Schwierigkeitsgrades (Abschnitte vorwiegend mathematischen Inhalts sind durch * gekennzeichnet) als auch zum Selbststudium. Dazu sind die Übungsaufgaben sowohl nach den benötigten mathematischen Kenntnissen als auch nach ihrem Schwierigkeitsgrad sorgfältig klassifiziert.
Auch dieser Band ist durch einen Anhang, in dem die in den vorangegangenen Bänden eingeführten Bezeichnungen erklärt werden, weitgehend unabhängig benutzbar. *Gerhard Maeß*

Zbl 1178.68372
Knuth, Donald E.
The art of computer programming. Volume 4, Fascicle 0-4. Fascicle 0: Introduction to combinatorial algorithms and Boolean functions. Fascicle 1: Bitwise tricks & techniques, binary decision diagrams. Fascicle 2: Generating all tuples and permutations. Fascicle 3: Generating all combinations and partitions. Fascicle 4: Generating all trees. History of combinatorial generation.
Upper Saddle River, NJ: Addison-Wesley (ISBN 978-0-3216-3713-0/pbk/set; 978-0-321-53496-5/Fascicle 0; 978-0-321-58050-4/Fascicle 1; 978-0-201-85393-3/Fascicle 2; 978-0-201-85394-0/Fascicle 3; 978-0-321-33570-8/Fascicle 4). Fas-

cicle 0, xii, 216 p.; Fascicle 1, viii, 260 p.; Fascicle 2, v, 129 p.; Fascicle 3, iv, 150 p.; Fascicle 4, vi, 120 p. EUR 74.22; £ 65.99 (2009).

Donald E. Knuth started an effort to write a book on compilers in 1962, but soon realized that its scope would have to be much wider. This led to *Fundamental Algorithms*, the first volume of *The Art of Computer Programming*, as published in 1968. Volume 2 on *Seminumerical Algorithms* and Volume 3 on *Searching and Sorting* followed in 1969 and 1973, respectively. All of them have been revised since, with the current versions published in the later 1990's.

I believe most of the computer scientists will agree that this collection of three volumes provides "a definitive description of classical computer science", as Addison-Wesley, its publisher, claims. Indeed, many of us have been growing up professionally with the nutrition as provided by these books, and we are certainly looking forward to seeing continuous efforts made towards completing this ambitious project, which is supposed to contain seven volumes [see *D. E. Knuth's* website, The Art of Computer Programming (TAOCP), available at http://www-cs-faculty.stanford.edu/~knuth/taocp.html]. The subject of Volume 4 is *combinatorial algorithms*, which deal with structured objects, thus often difficult to design, construct and analyze. On the other hand, most of the algorithms that we have to apply to real applications, e.g., graph algorithms, fall into this category, thus the need. Since the author feels that "the task of completing Volume 4 will take many years" and he "can't wait for people to begin reading what" he has "written so far and provide valuable feedback", Knuth has made available some segments of Volume 4, referred to as *fascicles*, initially online at the web page devoted to these books [loc. cit.], which were then published by Addison-Wesley between 2005 and 2009.

This collection of five fascicles, containing more than 870 pages, constitutes Section 7.1 and Section 7.2.1, part of Chapter 7 on *Combinatorial Searching*. This is the subject of our review, a task similar to writing a review for the Bible.

After introducing a rich set of exemplary combinatorial structures, Knuth focuses on the combinatorial structure generation problem in Section 7.2.1. To let readers better understand his intention, he makes a distinction among several related tasks at the very beginning of the subject [see p. 1 of Fascicle 2]. For example, by "enumerating" all the permutations of 1, 2, and 3, he means the number 6, i.e., the number of all the permutations for these three symbols; by "listing" these permutations, he means the task of making such a list, i.e., 123, 132, 213, 231, 312, and 321. On the other hand, by "generating" all the permutations of these three numbers", he means "to have them present momentarily in some data structure, so that a program can examine each permutation one at a time."

The above enumeration problem has been thoroughly investigated in the literature, such as [R. *Stanley*, Enumerative combinatorics. Vol. 1. 2nd ed. Cambridge Studies in Advanced Mathematics. 49. Cambridge: Cambridge

University Press (1997; Zbl 0889.05001)], [*P. Flajolet* and *R. Sedgewick*, Analytic combinatorics. Cambridge: Cambridge University Press (2009; Zbl 1165.05001)], while the listing problem is clearly subsumed by the generating problem. On the other hand, although this generating problem has been discussed in some of the applied combinatoric studies, e.g., [*A. Tucker*, Applied combinatorics. 5th ed. Hoboken, NJ: Wiley (2007)], it gets an encyclopedic treatment in this set of fascicles, where Knuth studies the generation of all the tuples, permutations, combinations, integer partitions, set partitions and trees, with the focus being placed on methods, i.e., algorithms, models and certainly mathematical ideas behind the scene, lots of them.

We do need such a generating mechanism often. For example, when looking for the number of all the shortest paths between two nodes in an arrangement graph [*K. Day* and *A. Tripathi*, "Arrangement graphs: a class of generalized star graphs", Inf. Process. Lett. 42, No. 5, 235–241 (1992; Zbl 772.68005)], I have a need to construct all the partitions of some g cycles in $n - k$ classes $T_1, \ldots, T_{n-k}$, by

(1) forming a tuple $(t_1, t_2, \ldots, t_{n-k})$ such that for all $i \in [1, n - k]$, $t_i \geq 0$, and $t_1 + t_2 + \cdots + t_{n-k} = g$; and

(2) for any such a tuple $(t_1, t_2, \ldots, t_{n-k})$, selecting t_1 cycles out of a total of g_l cycles in $\binom{g_l}{t_1}$ ways and place them in T_1, selecting t_2 cycles out of the remaining $g_l - t_1$ cycles in $\binom{g_l - t_1}{t_2}$ ways and place them in T_2, ..., and finally selecting the remaining t_{n-k} cycles in exactly one way and place them in T_{n-k}.

It is clear that Step 1 is indirectly involved with an integer partition generation problem [Algorithm H, p. 38 of Fascicle 3], while Step 2 is directly involved with a combination generation problem [Algorithm L, p. 4 in Fascicle 3]. In both cases, it is not enough to merely enumerate such partitions, but there is no need to explicitly list all such partitions, either. What we need here is to exhaustively come up with each and every such partition so that some succeeding construction(s) and/or operation(s) can be carried out. This is indeed what the "generating" operation is intended to do in this set of fascicles.

Since combinatorial algorithms are notoriously time-consuming, and a binary system can not only encode all the objects, but also potentially lead to a significant speed up of some of the combinatorial algorithms, Knuth discusses in detail the binary system in Section 7.1, including Boolean functions and their efficient evaluation, algorithms with heavy use of Boolean operations, i.e., bitwise operators such as those used in system programming, and binary decision diagrams, an efficient data structure to represent and manipulate Boolean functions.

This subsection ends with a historic review of combinatorial generation with an international scope.

Although we have to wait for many other interesting subjects to come out later, e.g., backtracking in Section 7.2.2, the shortest path problem in Section 7.3, graph and network algorithms in Sections 7.4 and 7.5, optimization-related discussion in Sections 7.7 to 7.9, and recursion in Section 8, let's do enjoy what we have here in front of us, another typical "Knuth book": technically correct, thoroughly researched, richly cited, elegantly written, and actually quite readable, *if* you make a serious effort.

By the way, if you have a good deal of time to spare, you can certainly try some of the exercises, 366 of them for just Section 7.1.1; and if you can't figure out an answer for some of the problems (the problem as given in Exercise 133 of Section 7.1.1 takes an MIT Ph.D. thesis to solve), you can check out the answers. In fact, Knuth has provided an answer, either a hint, a reference, or a complete solution, to almost all of the exercises as contained in these fascicles, with Exercise 92 of Section 7.1.1 being one of the few exceptions. As a result, e.g., out of the 260 pages of Fascicle 1, the segment on *Answers to Exercises* takes 95 pages.

Again, I would whole-heartedly recommend this book to my colleagues who have some time to spare, and would like to get into one of the greatest minds in Computer Science. *Zhizhang Shen (Plymouth/New Hampshire)*

Zbl 223.14022
Mumford, David
Abelian varieties.
Tata Institute of Fundamental Research Studies in Mathematics. 5.
London: Oxford University Press. VIII, 242 p. £5.00 net. (1970).

Author profile:

Mumford, David B.

Spellings: Mumford, David [78] Mumford, D. [42] Mumford, David B. [1]
Author-Id: mumford.david-b
Publications: 121 including 40 Book(s) and 55 Journal Article(s)

MSC 2010

65	14 · Algebraic geometry
13	68 · Computer science
11	00 · General mathematics
9	01 · History; biography
7	32 · Functions of several complex variables and analytic spaces

more ...

Journals

8	American Journal of Mathematics
4	Inventiones Mathematicae
4	Notices of the American Mathematical Society
2	Atti della Accademia Nazionale dei Lincei. Classe di Scienze Fisiche, Matematiche e Naturali. Serie IX. Rendiconti Lincei. Matematica e Applicazioni
2	Compositio Mathematica

more ...

Co-Authors

4	Michor, Peter W.
3	Fiath, Daniel E.
3	Gleason, Andrew M.
3	Gordon, Sheldon P.
3	Hughes-Hallett, Deborah

more ...

Publication Years

This is a remarkable book. The analytic and the algebraic theory of abelian varieties are brought together, which up to now did not happen before in one volume. The last 30 years it seems to be a sophisticated custom in algebraic geometry to hide geometric ideas and analytic motivations by wrapping them up into certainly useful, but often rather dry and complicated algebraic or functorial terminology. In this book the author uses a direct style, simplified notations and advanced methods. This sounds rather contradictory, but the result is a book which carries the reader immediately to the heart of the matter. An abelian variety is a complete irreducible group variety; the group structure is automatically commutative under these conditions; these varieties turned up naturally in the study of abelian integrals and their periods, which seems to explain the terminology. An abelian variety over the complex numbers is a complex torus, i.e. a compact connected complex Lie group, thus an analytic manifold of the form $\mathbb{C}^g/U$, where $U \cong \mathbb{Z}^{2g}$ is a lattice in $\mathbb{C}^g$; conversely such an analytic manifold is an algebraic manifold, and hence an abelian variety, if and only if it can be embedded into a projective space. Abelian varieties not only turn up in analytic theories but also in algebraic number theory and in various aspects of algebraic geometry. A connected group variety is an extension of an abelian variety by a linear group; thus the study of group varieties naturally splits into two. In contrast with the theory of linear algebraic groups, for abelian varieties the group theory is easy, but the geometry is complicated. In the first chapter (in less than 40 pages) the analytic theory is developed. Special attention is paid to the computation of cohomology groups, results which further in the book will find their algebraic analogues. Anyone who has struggled through the book of F. Conforto [Abelsche Funktionen und algebraische Geometrie (Berlin) (1956; Zbl 74.36601)], or who has studied A. Weil' s book [Introduction á l'etude des varietes kähleriennes (Paris) (1958; Zbl 137.41103)] will admire this penetrating approach.

The second chapter, algebraic theory via varieties, sets the first example of the conviction of the author that sheaves and their cohomology yield important tools. The fact that multiplication by n on an abelian variety of dimension g has degree n^{2g}, first proved by A. Weil in the algebraic case, obvious in the analytic case, is proved here in a short way. One of the new ideas exposed in the book is a direct construction of the dual abelian variety, i.e. its Picard variety, namely by dividing out X by the kernel of the isogeny $\varphi_L : X \to \mathrm{Pic}(X)$ defined by a polarization. In order to be able to do so, one has to prove that $\mathrm{Ker}(\varphi_L)$ is an algebraic group (scheme), which is not difficult in characteristic zero, which is more subtle in characteristic $p > 0$ due to the fact that φ_L may be inseparable (and hence $\mathrm{Ker}(\varphi_L)$ non reduced). The third chapter, algebraic theory via schemes, centers around this technical, and certainly important point. The proof of the strengthened version of the theorem of the cube (pp. 89–93) sets a fine example in which way scheme theoretic methods supply the necessary infinitesimal techniques to match algebraic rigidity and geometric intuition. Thus direct methods prove that

$K(L) = \text{Ker}(\varphi_L)$ is an algebraic group scheme, and the construction of $\hat{x} = X/K(L)$ follows. The methods of descent are applied to give a short proof of the duality theorem: for any isogeny $f : X \to Y$, the kernel of the dual of f is the dual of the finite group scheme $\text{Ker}(f)$. The methods of sheaf cohomology are used to prove the vanishing theorem and the Riemann Roch theorem.

The last chapter deals with the endomorphism ring: the algebraic restrictions on $E = \text{Hom}(X, X)$ coming from geometric properties of X, a nice explanation of the different isogeny classes of a CM type abelian variety according to different embeddings of the field of complex multiplications into the complex numbers, and elliptic curves. The last two sections cover the theory of theta groups and the Riemann form of a line bundle (corollary: every abelian variety is isogenous to a principally polarized abelian variety) plus its relation with the analytic form. Thus the circle closes: the analytic results exposed in beginning and end of the book nicely explain and profit from their algebraic analogues. The printing and the way of displaying formulas is of high quality; we mention two misprints: an Page 196, last formula: read $\sum$ instead of $\prod$; an Page 217, line 5 from bottom, read "all" instead of "suitable". The author is somewhat brief an the point of references; for two reasons it seems desirable to have more: firstly, it is not clear the reader always knows his way to the literature for concepts used but not defined in this book (a few more references would be of great help to non specialists); secondly, an several points theorems are proved, which also can be found in earlier literature; for some readers it may be easier to read earlier versions, and it seems fair to others to include references to earlier proofs of nontrivial results. We admire the style of exposition. Instead of giving explicitly in full detail all arguments which would have resulted in a doubly sized unreadable book the author phrases his proofs in such a way that verification of the arguments can be worked out, an enjoyable and stimulating task. In this way the book favourably contrasts with some modern volumes in which the real issues get lost in too many details and expanded notations. Experience has shown that a student with basic knowledge in algebraic geometry and some guidance is able to enjoy the treasures hidden in these 240 pages. There are several new results. Some of them were "well-known" but never published, several of them are contributions of the author. We mention the clear style and improvements due to the efforts of C. P. Ramanujam when writing up the notes after the course given at the Tata Institute of Fundamental Research. We strongly recommed this book, if not to every mathematician, certainly to analysts, geometers and algebraists. Getting every detail straight can be a task of several weeks, but obtaining a feeling for the many aspects of this beautiful field certainly is possible after some enjoyable hours of reading.   *Frans Oort*

Zbl 1177.14001
Mumford, David
Abelian varieties. With appendices by C. P. Ramanujam and Yuri Manin.
New Delhi: Hindustan Book Agency/distrib. by American Mathematical
Society (AMS); Bombay: Tata Institute of Fundamental Research (ISBN
978-81-85931-86-9/hbk). xii, 263 p. $ 30.00 (2008).

Presented annually by the American Mathematical Society, the Leroy P. Steele
Prize for Mathematical Exposition is one of the highest distinctions in Mathematics. In this category, the 2007 AMS Steele Prize was awarded to David
B. Mumford in recognition of his pioneering and beautiful accounts of a
host of aspects of algebraic geometry, published over a period of more than
two decades. The prize citation emphasizes that, particularly in D. Mumford
"The red book of varieties and schemes" (1974; Zbl 945.14001); includes the
Michigan lectures (1974) on "Curves and their Jacobians". 2nd, expanded
ed. with contributions by Enrico Arbarello. Lect. Notes Math. 1358. Berlin:
Springer 306 (1999; Zbl 223.14022)] the classical heritage is beautifully intertwined with the modern aspects of algebraic geometry, in a masterful
way that sharply illuminates both. Recognizing also the series of Mumford's
other influential books published between 1965 and 1991, the prize citation
concludes: "All of these books are, and will remain for the foreseeable future,
classics to which the reader returns over and over".

The book under review is the reprint of one of these distinguished expository
writings of D. Mumford, namely a corrected, completely re-typeset printing
of his particularly emphasized classic "Abelian Varieties".

Based on series of lectures delivered in the winter term 1967–1968 at the Tata
Institute of Fundamental Research (Mumbai/Bombay, India) and written up
by the late C. P. Ramanujam, the first edition of this book appeared in 1970.
The second edition, with two appendices added by C. P. Ramanujam and by
Yu. I. Manin, was brought out in 1974 (Zbl 0326.14012), and the first reprint of
this enlarged second edition was made available in 1985, that is, more than
20 years ago (Zbl 583.14015).

Now, in response to continuing strong demand, the Tata Institute of Fundamental research has finally provided another reprint of the second edition of
Mumford's classic "Abelian Varieties". In the current new reprint, the well-tried text has been left entirely intact. However, some appropriate corrections
have been supplied by B. Conrad and by Ching-Li Chai, and the book appears now in modern LaTeX typesetting, with the original page numbers
indicated in the margins of the present edition.

Back in 1970, Mumford's book provided the first modern treatment of abelian
varieties. Without approaching the subject by using "the crutch of Jacobian
or Albanese varieties", as it was still done in S. Lang's book "Abelian Varieties" (1958; Zbl 98.13201) from 1958, Mumford systematically developed
the basic theory of abelian varieties both in its general abstract setting and in
the framework of algebraic schemes and sheaves à la A. Grothendieck and P.
Cartier, thereby clarifying the picture of the situation also in positive charac-

teristics. Moreover, in Mumford's book, both the analytic and the algebraic theory of abelian varieties were treated simultaneously for the first time, and for the last time so far, which represented another novelty of propelling impact.

As D. Mumford himself pointed out in the preface, this book covers roughly half of the material that would be appropriate for a reasonably complete treatment of the theory of abelian varieties. Such a discussion should also include Jacobians, abelian schemes, deformation theory and moduli problems, modular forms and the global structure of moduli spaces, Dieudonné, theory in characteristic p, rationality questions, and the arithmetic theory of abelian schemes. However, such a comprehensive text on abelian varieties still has to be written, even almost forty years after the appearance of Mumford's modern introductory text on the subject. Therefore, apart from the various recent excellent textbooks on complex tori and complex abelian varieties (Birkenhake-Lange, Kempf, Debarre, Polishchuk, and others) and from J. S. Milne's recent, just as excellent lecture notes "Abelian Varieties" (version of March 16, 2008, available from the author's web site http://www.jmilne.org/math/CourseNotes/math731.html), where many of the further-leading algebraic aspects are touched upon, Mumford's classic is now as before the definite standard text for entering the modern theory of abelian varieties. Studying Mumford's unique book means learning directly from one of the great masters in the field, who was awarded a Fields Medal in 1974 for his epoch-making work on algebraic curves, abelian varieties, and their classification theory. It is Mumford's mastery of expository writing, which stands out by its unequalled clarity, elegance, originality and depth, that makes all his books and papers timeless classics, in particular his masterpiece "Abelian Varieties".

As the current new edition of this book is an unaltered reprint of the second edition from 1974, we may refer to the review of the latter D. *Mumford* Abelian varieties. With appendices by C. P. Ramanujam and Yuri Manin. Tata Institute of Fundamental Research Studies in Mathematics. 5. London: Oxford University Press. X, 279 p. £2.95 (1974; Zbl 326.14012)] respecting the detailed contents. However, after so many decades, it might be worthwhile to recall that the text comprises four chapters and two appendices discussing the following topics:

Chapter I: Analytic Theory (complex tori, line bundles, and algebraizabilty of tori);

Chapter II: Algebraic theory via varieties (abelian varieties, cohomology and base change, the theorem of the cube, quotients of varieties by finite groups, the dual abelian variety in characteristic 0, and the complex case);

Chapter III: Algebraic theory via schemes (basic theory of group schemes, quotients by finite group schemes, the general dual abelian variety, duality for finite commutative group schemes, applications to abelian varieties, cohomology of line bundles, and very ample line bundles);

Chapter IV: Endomorphisms of Abelian Varieties and ℓ-adic Representations (étale coverings, structure of endomorphism rings of abelian varieties, Riemann forms, the Rosati involution, and Riemann forms via theta groups); Appendix I (by C. P. Ramanujam): The Theorem of Tate; Appendix II (by Yu. I. Manin): The Mordell-Weil theorem.

All in all, it is more than rewarding that D. Mumford's classic standard text "Abelian Varieties" has been made available again for new generations of students, teachers, and researchers in the field. In its modern typesetting and in its corrected version, this jewel in the mathematical literature has become even more attractive and valuable.   *Werner Kleinert (Berlin)*

Zbl 258.14011
Dieudonné, Jean A.; Carrell, James B.
Invariant theory, old and new.
New York-London: Academic Press. VIII, 85 p. $ 4.75 (1971).

This book is reprinted from an article of the authors [Adv. Math. 4, 1–80 (1970; Zbl 196.05802)] and has been reviewed by the senior author. Because it has now appeared in bock form making it more readily available, because of the renewed interest in the subject and because it is a well written, exciting introduction to invariant theary, it is worth another review. The preface gives a brief history of invariant theory which serves also as an outline of the book itself.

Author profile:

Dieudonné, Jean Alexandre

Spellings: Dieudonné, Jean [289] Dieudonné, J. [65] Dieudonné, Jean A. [11] Dieudonné, Jean Alexandre [1] Dieudonné, J.A. [1]
Author-Id: dieudonne.jean-alexandre
Publications: 367 including 81 Book(s) and 256 Journal Article(s)

MSC 2010

54	01 · History; biography
21	46 · Functional analysis
20	22 · Topological groups, Lie groups
17	00 · General mathematics
16	47 · Operator theory

more ...

Journals

45	Comptes Rendus de l'Académie des Sciences. Paris
12	Bulletin de la Société Mathématique de France
10	American Journal of Mathematics
9	Mathesis
8	Bulletin des Sciences Mathématiques. Deuxième Série

more ...

Co-Authors

5	Grothendieck, Alexander
3	Carrell, James B.
2	Tits, Jacques
1	Behnke, Henri
1	Bourbaki, Nicolas

more ...

Publication Years

Chapter 1 introduces the terminology and ends with the important restitution process and the reduction from the multilinear to the linear case. Stated briefly, we have a group G acting an a set E and we wish to describe the set of invariant elements $E^G = \{e \in E : s \cdot e = e$ for all $s \in G\}$. Often the space E is a vector space and the action of G is required to be linear. Real geometric problems arise when G is an algebraic group defined over a field k, the set E is a k scheme and the action is a morphism of k-schemes. The question then is whether the space of orbits is again a k-scheme.

In Chapter 2, entitled: "Rational representations of the general linear group", the synthetic method of the 19th century is developed and used firstly to show that the general linear group $GL(n, k)$ defined over an algebraically closed field k of characteristic zero is completely reducible and then secondly to find all rational eharacters. Finally Aronhold's rule is developed so that all irreducible rational representations ran be obtained. [Compare this method with the more modern method found, for example, in the article of R. *Hartshorne*, Publ. Math., Inst. Hautes Étud. Sci. 29, 63–94 (1966; Zbl 173.49003) where Borel's fixed point theorem is used in an essential manner. See also J. *Dieudonné*, Sem. Bourbaki 1970/71, Lect. Notes Math. 244, 257–274 (1971; Zbl 231.01001) for a description of the f irst two chapters of this bock under review.]

Author profile:

Carrell, James B.

Spellings: Carrell, James B. [34] Carrell, J.B. [14] Carrell, Jim [1] Carrell, James [1] Carrell, J. [1]
Author-Id: carrell.james-b
Publications: 51 including 7 Book(s) and 28 Journal Article(s)

MSC 2010

39	**14** · Algebraic geometry
25	**32** · Functions of several complex variables and analytic spaces
10	**57** · Manifolds and cell complexes
9	**20** · Group theory and generalizations
7	**53** · Differential geometry

more ...

Journals

3	Inventiones Mathematicae
3	Mathematica Scandinavica
2	Compositio Mathematica
2	Journal of Differential Geometry
2	Mathematische Annalen

more ...

Co-Authors

8	Sommese, Andrew John
6	Akyildiz, Ersan
5	Lieberman, David I.
3	Dieudonné, Jean Alexandre
2	Kaveh, Kiumars

more ...

Publication Years

Chapter 3, "Post Hilbert invariant theory", the problem of the finite generation of the invariants of a finitely generated commutative k-algebra is discussed. A complete proof is given of the Hilbert-Nagata theorem: Suppose G is a group of automorphisms of the finitely generated k-algebra R subject to the two conditions; (i) the orbit of each element of R under G is contained in a finite dimensional vertor space over k, and (ii) each linear representation of G an a finite dimensional vector space over k which leaves invariant a hyperplane is equivalent to a representation of the same dimension fixing pointwise a line complementary to an invariant hyperplane. Then the ring of invariants R^G is finitely generated over k. This is related to Hilbert's 14th problem: Suppose $GL(n,k)$ actc in the usual manner on the ring of polynomials $k[X_1, \ldots, X_n]$. Let G be a subgroup of $GL(n,k)$. Is the ring $k[X_1, \ldots, X_n]^G$ a finitely generated k-algebra? It is easily seen that the action of G satisfies condition (i) above. If G is reductive, then condition (ii) is satisfied. However $M.$ *Nagata* [Notes by M. Pavaman Murthy. (Tata Institute of Fundamental Research Lectures on Mathematics and Physics. Mathematics. Vol. 31) Bombay: Tata Institute of Fundamental Research. 78 + III p. (1965; Zbl 182.54101)] has found a group G for which the fixed subring is not finitely generated. This counterexample to Hilbert's 14th problem is reported in the last section of this chapter with most of the details verifying its existence given.

The final Chapter 4, "Introduction to Hilbert-Mumford theory", is exactly as the title states. Whereas Chapter 3 is concerned with construction of quotients of affine schemes, the main concern of Chapter 4 is the construction of orbit spaces (or quotients) of projective varieties. This is a more difficult, not to say insoluble, problem since it must be solved first locally and then the local orbit spaces must be glued together coherently. This is perhaps the most interesting chapter because it leads the reader naturally to Mumford's recent revival of geometric invariant theory [*D. Mumford*, Geometric invariant theory, Berlin-Heidelberg-New York (1965; Zbl 147.39304)]. The chapter ends with several illustrative examples.

Finally there is an appendix, "A short digest of non-commutative algebra", which contains the theory of semi-simple algebras, including Maschke's theorem and Schur's commutation theorem, necessary for showing the complete reducibility of $GL(n,k)$. This material is used only in Chapter 2.

This book is an excellent introduction to the subject giving not only the important contributions of the last century, but also the most important contributions of the last decades, as well as an introduction to the recent developments due to Mumford. Recent successes of invariant theory have been made by Hochster, Eagon and Roberts. *M. Hochster* and *J. A. Eagon* have shown, using invariant theory, that the homogeneous coordinate ring for the Plücker embedding of the Grassmannian is Gorenstein [Am. J. Math. 93, 1020–1058 (1971; Zbl 244.13012)], while *M. Hochster* has shown that the ring of invariants of a torus is Cohen-Macaulay [Ann. Math. (2) 96, 318–337 (1972; Zbl 233.14010)]. Most recently *M. Hochster* and *Roberts* have shown that a linearly reductive affine linear group defined over a field k (of arbitrary characteristic) acting on a regular k algebra has a Cohen-Macaulay ring of invariants [Rings of invariants of reductive groups acting on regular rings are Cohen-Macaulay. Mimeographed. Univ. of Minnesota, Minneapolis, Minnesota (1973)]. *Robert M. Fossum (Champaign)*

Zbl 257.46078
Feferman, Charles Louis; Stein, Elias M.
H^p spaces of several variables.
Acta Math. 129, 137-193 (1972).

This paper is a major contribution to the study of H^p spaces, singular integrals, and harmonic analysis an $\mathbb{R}^n$. Classically the theory of H^p spaces arose from analytic function theory. H^p was defined as the space of analytic functions in the upper half plane with boundary values in $L^p(\mathbb{R})$. More recently this definition has been generalised to $\mathbb{R}^n$ by introducing generalised conjugate harmonic functions in $\mathbb{R}^{n+1}_+ = \{(x, t) : x \in \mathbb{R}^n, t > 0\}$. The authors present several intrinsic descriptions of H^p, of a real variable nature, not involving conjugate functions. These results greatly clarify the meaning of H^p, as well as throwing new light on the behaviour of convolution operators an L^p. The following is a summary of some of the main results. The first main result is the description of the dual of H^1. H^1 is the Banach space of all functions f in $L^1(\mathbb{R}^n)$ such that $R_j f \in L^1(\mathbb{R}^n)$, $j = 1, \ldots, n$, where R_j is the j-th Riesz transform. (In terms of Fourier transforms, $(R_j f)^\wedge(y) = y_j \hat{f}(y)/|y|$. When $n = 1$ the definition says that the Hilbert transform of f is in L^1, or equivalently $\int f(x)\, dx = 0$ and $f = g + \bar{h}$ where g and h are in the classical "analytic" H^1). The authors prove that the dual of H' is the space of all functionals of the form $\varphi \to \int f\varphi$ (suitably interpreted if $f\varphi \notin L^1$), where φ is a function of bounded mean oscillation (BMO), which means that there is a constant $C > 0$ (depending an φ) such that $\int_Q |f - f_Q| \leq C|Q|$ for any cube Q in $\mathbb{R}^n$, where $f_Q = |Q|^{-1} \int_Q f$. The proof of this remarkable and deep result depends on an inequality of Carleson and some clever manipulations of Littlewood Paley functions. The hard part is to show that a BMO function defines a bounded functional on H^1. The essential difficulties are already present in the case $n = 1$. This duality leads to a new approach to convolution operators which brings out the usefulness of BMO as a substitute for L^∞. If T is a convolution operator (i.e. $Tf = K * f$ for some distribution K an $\mathbb{R}^n$) mapping L^∞ into BMO, then the authors show, using the duality and a description of L^p related to BMO, that T maps H^1 into H^1, BMO into BMO, and L^p into L^p for $1 < p < \infty$. The condition that T map L^∞ into BMO is relatively easy to verify for singular integral operators of Calderón-Zygmund type. A more refined version of this result, involving Calderón's complex method of interpolation, enables the authors to prove new results on L^p multipliers. The authors then turn to H^p spaces for general p ($0 < p < \infty$). They define these first as spaces of harmonic functions an $\mathbb{R}^{n+1}_+$, without reference to boundary values. Specifically, a harmonic function u_0 is in H^p if there exist harmonic functions

128

Author profile:

Fefferman, Charles Louis

Spellings: Fefferman, Charles [85] Fefferman, C. [36] Fefferman, Charles L. [19] Fefferman, C.L. [7] Fefferman, Charles Louis [3] Fefferman, Ch. [1]
Author-Id: fefferman.charles-louis
Publications: 151 including 1 Book(s) and 124 Journal Article(s)

MSC 2010

47	**35** · Partial differential equations (PDE)	
30	**81** · Quantum Theory	
21	**32** · Functions of several complex variables and analytic spaces	
19	**58** · Global analysis, analysis on manifolds	
18	**42** · Fourier analysis	

more ...

Journals

18	Revista Matemática Iberoamericana
13	Annals of Mathematics. Second Series
12	Proceedings of the National Academy of Sciences of the United States of America
11	Advances in Mathematics
7	Communications in Mathematical Physics

more ...

Co-Authors

19	Seco, Luis A.
10	Phong, Duong Hong
9	Cordoba, Diego
8	Donnelly, Harold
7	Córdoba, Antonio Juan

more ...

Publication Years

$u_1, \ldots, u_n$ satisfying

$$\partial u_j / \partial x_i = 0, \quad \sum_{i=0}^{n} \partial u_i / \partial x_i \quad \text{and} \quad \sup_{t>0} \int_{\mathbb{R}^n} |u(x,t)|^p \, dx < \infty,$$

where $|u|^2 = \sum_{i=0}^{n} |u_i|^2$. This definition is appropriate if $p > (n-1)/n$. In general a more elaborate version (here omitted) is needed (the point is that $|u|^p$ is subharmonic only if $p \geq (n-1)/n$). The main result is as follows: let u be harmonic in $\mathbb{R}_+^{n+1}$ and define $u^*(x) = \sup_t |u(x,t)|$. Then $u \in H^p$ if and only if $u^* \in L^p$. (For $n = 1$ this was proved by Burkholder, Gundy and Silverstein in 1971). Finally the authors consider boundary values. If $u \in H^p$ then $u(x,t) \to f(x)$ as $t \to 0$, in the distribution sense, where f is a tempered distribution on $\mathbb{R}^n$. Denote the set of such f also by H^p. Then $H^p = L^p$ for $p > 1$ and for $p = 1$ this definition is consistent with the earlier one. The last result above characterises H^p in terms of Poisson integrals (as u is the Poisson integral of f). The authors show that the Poisson kernel can be replaced by any smooth approximate identity – more precisely, fix a smooth function φ on $\mathbb{R}^n$, decreasing rapidly at ∞, with $\int \varphi = 1$. Put $\varphi_t(x) = t^{-n}\varphi(x/t)$, and for any tempered disribution f write $f^*(x) = \sup_{t>0} |\varphi_t * f(x)|$. Then (for $0 < p < \infty$) $f \in H^p$ if and only if $f^* \in L^p$. ("Non-tangential" versions of this and the preceding result are also given). This result implies for example that one could define H^p in terms of the wave equation rather than Laplace's and get the same space of functions on $\mathbb{R}^n$. The paper concludes with a proof that certain singular integral operators map H^p to itself.   *Alexander M. Davie*

Author profile:

Stein, Elias M.

Spellings: Stein, Elias M. [211]
Author-Id: stein.elias-m
Publications: 211 including 16 Book(s) and 168 Journal Article(s)

MSC 2010

76	**42** · Fourier analysis
41	**43** · Abstract harmonic analysis
31	**35** · Partial differential equations (PDE)
26	**22** · Topological groups, Lie groups
23	**32** · Functions of several complex variables and analytic spaces

more ...

Journals

21	Proceedings of the National Academy of Sciences of the United States of America
16	Annals of Mathematics. Second Series
11	Acta Mathematica
9	Inventiones Mathematicae
8	American Journal of Mathematics

more ...

Co-Authors

20	Nagel, Alexander
19	Wainger, Stephen
15	Phong, Duong Hong
11	Knapp, Anthony W.
10	Ricci, Fulvio

more ...

Publication Years

Zbl 292.18004
Quillen, Daniel
Higher algebraic K-theory. I.
Algebr. K-Theory I, Proc. Conf. Battelle Inst. 1972, Lect. Notes Math. 341,
85-147 (1973).

Author profile:

Quillen, Daniel G.

Spellings: Quillen, Daniel [35] Quillen, D. [7] Quillen, D.G. [6] Quillen, Daniel G. [3]
Author-Id: quillen.daniel-g
Publications: 51 including 2 Book(s) and 38 Journal Article(s)

MSC 2010

16	**55** · Algebraic topology
13	**18** · Category theory, homological algebra
10	**19** · K-theory
8	**20** · Group theory and generalizations
7	**16** · Associative rings and algebras

more ...

Journals

8	Topology
3	Annals of Mathematics. Second Series
3	Comptes Rendus de l'Académie des Sciences. Série I
3	Inventiones Mathematicae
2	Advances in Mathematics

more ...

Co-Authors

6	Cuntz, Joachim
3	Guillemin, Victor W.
2	Loday, Jean-Louis
2	Sternberg, Shlomo
1	Bousfield, A.K.

more ...

Publication Years

In this seminal paper, the author defines higher algebraic K-groups for certain additive categories and generalizes to the groups most of the classical techniques used to study the Grothendieck group K_0. Given a full additive subcategory M of an abelian category $\mathcal{A}$ which is closed under extensions in $\mathcal{A}$ define the category $Q(M)$ to have the same objects as M, with a morphism $M \to M$ being an isomorphism of M' with a subquotient M_i/M_o of M, where M_o and M/M_i are in M. If the isomorphism classes of objects of M form a set, the geometric realization of the nerve of $Q(M)$ is called the classifying space, $BQ(M)$; it is determined up to homotopy equivalence. By definition, $K_i(M) = \pi_{i+1}(BQ(M), 0)$.
The first section of this paper investigates the homotopy-theoretic properties of these classifying spaces and the maps induced by functors an the underlying categories. Section 2 contains the definition and elementary properties of the K-groups, including basic exact sequences. Section 3–5 are devoted

to proofs of the exactness, resolution, devissage and localization theorems which generalize well-known techniques for studying K_0 and K_1 and provide a first justification for the definition offered in §2.

The second part of this paper, §§6–8, applies the general theory to rings and schemes. For a ring (resp. noetherian ring) A, $K_i(A)$ (resp. $K'_i(A)$) are the K-groups of the category of finitely generated projective (resp. finitely generated) A-modules. Among the important results are:

(1) $K_j(A) \xrightarrow{\approx} K'_j(A)$ is regular noetherian.

(2) $K'_i(A) \approx K'_i(A[r])$; same for K_i if A is regular.

(3) $K'_i(A[t, t^{-1}]) \approx K'_i(A) \oplus K'_{i-1}(A)$; same for K_i if A is regular.

For a scheme (resp. noetherian scheme) X, $K_i(X)$ (resp. $K'_i(X)$) are the K-groups of the category of vector bundles (resp. coherent sheaves) an X. Filtering the category of coherent sheaves by codimension of support yields a spectral sequence

$$E_i^{pq} = \coprod_{\mathrm{cod}(x)=p} K_{-p-q}(k(x)) \Rightarrow K'_n(X).$$

When X is regular and of finite type over a field, this leads to a proof of Bloch's formula: $CH^p(X) = H^p(X, K_p(O_x))$, where $CH^p(X)$ is the group of codimension p cycles an X modulo linear equivalence.

This paper contains proofs of all results announced in [Higher K-theory for categories with exact sequences, to appear in the Proceedings of the June 1972 Oxford Symposium "New developments in topology"] except for the fact that the groups $K_i(A)$ introduced here agree with those defined via the $BGL(A)^+$ construction of [the author, Actes Congr. internat. Math. 1970, 2, 47–51 (1971; Zbl 225.18011)]. *Michael R. Stein*

Zbl 292.10035
Bombieri, Enrico
The large sieve in analytic number theory.
Astérisque 18, 1-87 (1974).

In this monograph the author presents a concise and beautiful account of the Large Sieve and some of its important applications to analytic number theory, in particular, the density theorems, the distribution of primes in arithmetical progressions and the connection with the small sieve of Brun-Selberg. The method of the Large Sieve, which was invented by *Yu. V. Linnik* [C. R. (Dokl.) Acad. Sci. URSS, n. Ser. 30, 292–294 (1941; Zbl 24.29302)] and has later been developed by A. Renyi, M. B. Barban, A. I. Vinogradov, K. F. Roth, E. Bombieri, P. X. Gallagher and others, is, as is widely recognized, one of the most powerful tools in multiplicative number theory. The booklet consists of Introduction, Notes in twelve paragraphs, Bibliography and Summary in English. The twelve paragraphs are: §0. Preliminaires. Notations and a definition of a sieve; §1. Quelques exemples. Le crible de Linnik et Renyi. The theorem of Linnik an the least quadratic non-residue (mod p), and Renyi's formulation of the large sieve; §2. La forme analytique additive du grand crible. A formulation of the large sieve due to Roth and Bombieri; §3. Applications arithmetiques. Le crible de Selberg (I); §4. La forme multiplicative du grand crible. The large sieve inequality containing Dirichlet's residue characters; §5. La forme analytique multiplicative du grand crible. The so-called hybrid sieve of Gallagher and some results due to M. Forti and C. Viola; §6. Applications. Le theoreme de Linnik. A density theorem for zeros of Dirichlet L-functions and the famous theorem of Linnik an the least prime in an arithmetic progression; §7. Applications. Le théorème des nombres premiers dans les progressions arithmétiques. A modification of the simplified proof by Gallagher of the Bombieri-Vinogradov theorem; §8. Le crible de Selberg (II). A generalization of the classical sieve of Selberg; §9. Application du crible de Selberg. A proof of the result that there are infinitely many primes p such that $p + 2$ has at most 4 prime factors; §10. Théorèmes de densité; §11. Notes Bibliographiques. The results demonstrated are not always the strongest ones known at present, but the exposition is an the whole very clear and readable.   *Saburo Uchiyama*

Author profile:

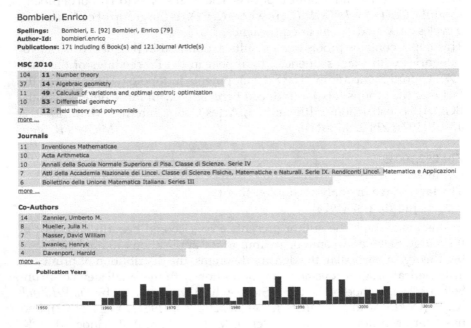

Bombieri, Enrico

Spellings: Bombieri, E. [92] Bombieri, Enrico [79]
Author-Id: bombieri.enrico
Publications: 171 including 6 Book(s) and 121 Journal Article(s)

MSC 2010

104	11 · Number theory
37	14 · Algebraic geometry
11	49 · Calculus of variations and optimal control; optimization
10	53 · Differential geometry
7	12 · Field theory and polynomials

more ...

Journals

11	Inventiones Mathematicae
10	Acta Arithmetica
10	Annali della Scuola Normale Superiore di Pisa. Classe di Scienze. Serie IV
7	Atti della Accademia Nazionale dei Lincei. Classe di Scienze Fisiche, Matematiche e Naturali. Serie IX. Rendiconti Lincei. Matematica e Applicazioni
6	Bollettino della Unione Matematica Italiana. Series III

more ...

Co-Authors

14	Zannier, Umberto M.
8	Mueller, Julia H.
7	Masser, David William
5	Iwaniec, Henryk
4	Davenport, Harold

more ...

Publication Years

| 1950 | 1960 | 1970 | 1980 | 1990 | 2000 | 2010 |

Zbl 351.92021
Li, Tien-Yien; Yorke, James A.
Period three implies chaos.
Am. Math. Mon. 82, 985-992 (1975).

Let F be a continuous function of an interval J into itself. The period of a point in J is the least integer $k > 1$ for which $F^k(p) = p$. If p has period 3 then the relation $F^3(q) \leq q < F(q) < F^2(q)$ (or its reverse) is satisfied for q one of the points p, $F(p)$, or $F^2(p)$. The title of the paper derives from the theorem that if some point q in J has this Sysiphutian fearure, "two steps forward, one giant step back", then F has periodic points of every period $K = 1, 2, 3, \ldots$. Moreover, J contains an uncountable subset S devoid of asymptotically periodic points, such that

Author profile:

Li, Tien-Yien

Spellings: Li, T.Y. [53] Li, Tien-Yien [32] Li, T. [21] Li, T.-Y. [5] Li, Tien Yien [3] Li, T. Y. [2]
Author-Id: li.tien-yien
Publications: 116 including 1 Book(s) and 97 Journal Article(s)

MSC 2010

65	**65** · Numerical analysis
10	**12** · Field theory and polynomials
10	**26** · Real functions
9	**30** · Functions of a complex variable
8	**68** · Computer science

more ...

Journals

8	SIAM Journal on Numerical Analysis
5	Mathematics of Computation
5	Nonlinear Analysis. Theory, Methods & Applications. Series A: Theory and Methods
5	SIAM Journal on Scientific Computing
4	Numerische Mathematik

more ...

Co-Authors

17	Yorke, James A.
10	Wang, Xiaoshen
9	Zeng, Zhonggang
6	Sauer, Tim D.
5	Ding, Jiu

more ...

Publication Years

$$0 = \liminf |F^n(q) - F^n(r)| < \limsup |F^n(q) - F^n(r)|$$

for all $q \neq r$ in S. (a point is asymptotically periodic if $\lim |F^n(p) - F^n(q)| = 0$ for some periodic point p.) The proof is eminently accessible to the nonspecialist

134

Author profile:

Yorke, James A.

Spellings: Yorke, James A. [185] Yorke, J.A. [42] Yorke, J. [4] Yorke, James [3] Yorke, J. A. [1]
Author-Id: yorke.james-a
Publications: 233 including 7 Book(s) and 185 Journal Article(s)

MSC 2010

132	**37** · Dynamical systems and ergodic theory
49	**34** · Ordinary differential equations (ODE)
25	**65** · Numerical analysis
22	**28** · Measure and integration
17	**54** · General topology

more ...

Journals

27	Physica D
17	Journal of Differential Equations
13	Ergodic Theory and Dynamical Systems
13	Transactions of the American Mathematical Society
8	Bulletin (New Series) of the American Mathematical Society

more ...

Co-Authors

26	Grebogi, Celso
25	Ott, Edward
16	Kennedy, Judy A.
16	Sauer, Tim D.
15	Alligood, Kathleen T.

more ...

Publication Years

| 1960 | 1970 | 1980 | 1990 | 2000 | 2010 |

and is therefore of interest to anyone modeling the evolution of a single population parameter by a first order difference equation.

The authors compare the logistic $x_{n+1} = F(x_n) = rx_n(1 - x_n/K)$ with a model of which, by contrast, $|dF(x)/dx| > 1$ wherever the derivative exists. For such a system no periodic point is stable, in the sense that $|F^k(q) - p| < |q - p|$ for all q in a neigborhood of a periodic point p of k. A brief survey of a theorem motivated by ergodic theory completes this fascinating paper. *G.K. Francis*

Zbl 343.46042
Connes, Alain
Classification of injective factors. Cases $\text{II}_1, \text{II}_\infty, \text{III}_\lambda, \lambda \neq 1.$
Ann. Math. (2) 104, 73-115 (1976).

The paper contains definitive results an hyperfiniteness and injectivity of von Neumann algebras, which give the solutions of many important problems in the theory of operator algebras. Let N be a von Neumann algebra on a Hilbert space H and $B(H)$ the algebra of all bounded linear operators in H. N is said to be injective if there is a projection of norm one of $B(H)$ to N or equivalently if, for a C^* algebra A and its C^*-subalgebra B, any completely

Author profile:

Connes, Alain

Spellings: Connes, Alain [161] Connes, A. [61]
Author-Id: connes.alain
Publications: 222 including 19 Book(s) and 133 Journal Article(s)

MSC 2010

132	**46** · Functional analysis
89	**58** · Global analysis, analysis on manifolds
60	**81** · Quantum Theory
25	**11** · Number theory
23	**57** · Manifolds and cell complexes

more ...

Journals

16	Communications in Mathematical Physics
13	Comptes Rendus de l'Académie des Sciences. Série I
9	Letters in Mathematical Physics
6	Journal of Functional Analysis
6	Journal of Geometry and Physics

more ...

Co-Authors

20	Moscovici, Henri
14	Marcolli, Matilde
9	Chamseddine, Ali H.
9	Kreimer, Dirk
7	Consani, Caterina

more ...

Publication Years

positive map of B into N has a completely positive extension to A [*J. Hakeda* and the reviewer, Tôhoku math. J., II. Ser. 19, 315–323 (1967; Zbl 175.14201); *E. Effros* and *C. Lanee*, Tensor products of operator algebras, to appear in Advances Math.]. The algebra N is also said to be semidiscrete if the identity map $N \to N$ is approximated in σ-weak topology by a net of completely positive maps of finite rank. The author's main result asserts that for a factor N of type II_1 in a separable Hilbert space the notions of injectivity and semidiscreteness are equivalent to the hyperfiniteness of N, the weak closure of an ascending sequence of matrix algebras (results are stated in separated theorems). He also proved further equivalence of these properties to those of the property P by *J. T. Schwartz* [Commun. Pure Appl. Math. 16, 19–26 (1963; Zbl 131.33201)] and the property Γ [*F. J. Murray* and *J. von Neumann*, Ann. Math. (2) 44, 716–808 (1943; Zbl 60.26903)]. Thus, as natural consequences of these results one knows that up to isomorphisms there is only one injective factor of type II_1, a hyperfinite factor and the hyperfinite factor of type II_∞ is unique. It is also now clear that all subfactors of a hyperfinite factor R of type III_1 are isomorphic to R or finite dimensional. The equivalences of those properties are further shown to be valid for any factor in a separable Hilbert spare. Besides these remarkable consequences, the result implies the following answer to the conjecture by Kadison and Singer; any representation of a solvable separable locally compact group or a connected locally compact separable group in a Hilbert space generates a hyperfinite von Neumann algebra. The paper also contains characterizations of an automorphism which lies in the closure of the inner automorphism group, $\operatorname{Int} N$, for a factor of type II_1. *J. Tomiyama*

Zbl 412.65058
Gottlieb, David; Orszag, Steven A.
Numerical analysis of spectral methods: Theory and applications.
CBMS-NSF Regional Conference Series in Applied Mathematics. 26.
Philadelphia, Pa.: SIAM, Society for Industrial and Applied Mathematics.
v, 172 p. $ 12.25 (1977).

In der Monographie werden Galerkin-, Tau- und Collocationsverfahren für Anfangsrandwertaufgaben untersucht. Die Ansatzfunktionen für die vielen eindimensionalen Testprobleme hyperbolischer und parabolischer Art sind Eigenfunktionen von Sturm-Liouville-Aufgaben. Die Konvergenzuntersuchungen für die entstehendem semidiskreten Probleme werden für eindimensionale, lineare und autonome Anfangsrandwertaufgaben durchgeführt. Hierbei wird ausführlich dargestellt, daß für den Spezialfall semibeschränkter Operatoren der dem Äquivalenzsatz von Lax zugrundegelegte Stabilitätsbegriff herangezogen werden kann, im allgemeinen Fall aber die schwächere algebraische Stabilität als sachgemäßer Stabilitätsbegriff für Spektralmethoden verwendet werden muss. Die schwächere Stabilität wird durch stärkere Konsistenzbedingungen kompensiert.

Für die zeitliche Diskretisierung der semidiskreten Probleme werden explizite und implizite Einschrittverfahren vorgeschlagen. Die abschließenden Kapitel behandelt die Implementierung der Spektralmethoden und den numerischen Vergleich mit anderen Verfahren, insbesondere Differenzenverfahren. Für das Verständnis der Monographie muß der Leser gute Grundkenntnisse bezüglich der numerischen Behandlung von Anfangswertaufgaben besitzen. Den Autoren ist es insbesondere gelungen, ihre großen praktischen Erfahrungen weiterzugeben.

Die numerischen Probleme bei der konkreten Problemlösung von Anfangsrandwertaufgaben, wie z. B. Gibbs-Phänomen, Phasenfehlern und Reflexion an Rändern werden ausführlich dargestellt. Die Monographie ist daher für solche Leser besonders empfehlenswert, die nicht primär an theoretischen Ergebnissen interessiert sind, sondern die Spektralmethode auf aktuelle Aufgaben anwenden wollen. *Horst Kreth*

Author profiles:

Gottlieb, David I.

Spellings: Gottlieb, David [87] Gottlieb, David I. [1]
Author-Id: gottlieb.david-i
Publications: 88 including 3 Book(s) and 72 Journal Article(s)

MSC 2010

59	65 · Numerical analysis
45	35 · Partial differential equations (PDE)
27	76 · Fluid mechanics
10	42 · Fourier analysis
5	33 · Special functions

more ...

Journals

16	Journal of Computational Physics
11	SIAM Journal on Numerical Analysis
8	Journal of Scientific Computing
8	Mathematics of Computation
6	Applied Numerical Mathematics

more ...

Co-Authors

16	Abarbanel, Saul S.
14	Shu, Chi-Wang
11	Carpenter, Mark H.
10	Turkel, Eli L.
7	Don, Wai Sun

more ...

Publication Years

Orszag, Steven A.

Spellings: Orszag, Steven A. [104] Orszag, S.A. [35] Orszag, Steven [2] Orszag, S. [2] Orszag, S. A. [1]
Author-Id: orszag.steven-a
Publications: 144 including 8 Book(s) and 107 Journal Article(s)

MSC 2010

117	76 · Fluid mechanics
32	65 · Numerical analysis
15	35 · Partial differential equations (PDE)
9	82 · Statistical mechanics, structure of matter
8	80 · Classical thermodynamics, heat transfer

more ...

Journals

29	Journal of Scientific Computing
28	Journal of Fluid Mechanics
12	Physics of Fluids
11	Journal of Computational Physics
5	Studies in Applied Mathematics

more ...

Co-Authors

16	Yakhot, Victor
11	Staroselsky, Ilya
9	Meiron, Daniel I.
8	She, Zhen-Su
7	Goldhirsch, Isaac

more ...

Publication Years

138

Zbl 418.05028
Lovász, László
Kneser's conjecture, chromatic number, and homotopy.
J. Comb. Theory, Ser. A 25, 319-324 (1978).

Author profile:

The author proves a result on the simplicial complex formed by the neighborhoods of points of a graph and applies that to obtain an elegant proof of the Kneser conjecture of 1955 asserting that if the n-subsets of a $(2n + k)$-element set are split into $k + 1$ classes, then one of the classes will contain two disjoint n-subsets. The proof depends on the theorem of K. Borsuk [Fundam. Math. 20, 177–190 (1933; Zbl 0006.42403; JFM 59.0560.01)] that if the k-dimensional unit sphere is covered by $k + 1$ closed sets, then one of these contains two antipodal points. *Carsten Thomassen (Lyngby)*

Zbl 426.60069

Stroock, Daniel W.; Varadhan, S.R.Srinivasa

Multidimensional diffusion processes.

Grundlehren der mathematischen Wissenschaften. 233.

Berlin, Heidelberg, New York: Springer-Verlag. XII, 338 p.

DM 70.00; $ 38.50 (1979).

This book develops the martingale approach to Markov diffusion processes, along the lines initiated by the same authors in two papers that appeared in 1969. This approach has been since widely used to construct and handle various types of Markov processes. The authors have chosen to cover the case of diffusion processes in $\mathbb{R}^d$, in a very complete way.

Author profile:

Stroock, Daniel W.

Spellings: Stroock, Daniel W. [132] Stroock, D. [15] Stroock, D.W. [12] Stroock, Daniel [8]
Author-Id: stroock.daniel-w
Publications: 167 including 23 Book(s) and 94 Journal Article(s)

MSC 2010

122	60 · Probability theory and stochastic processes
32	35 · Partial differential equations (PDE)
27	58 · Global analysis, analysis on manifolds
10	28 · Measure and integration
10	82 · Statistical mechanics, structure of matter

more ...

Journals

14	Journal of Functional Analysis
10	Communications on Pure and Applied Mathematics
5	Communications in Mathematical Physics
5	Zeitschrift für Wahrscheinlichkeitstheorie und Verwandte Gebiete
4	The Annals of Probability

more ...

Co-Authors

17	Holley, Richard A.
17	Varadhan, Srinivasa R.S.
11	Kusuoka, Shigeo
6	Jerison, David S.
5	Deuschel, Jean-Dominique

more ...

Publication Years

1960   1970   1980   1990   2000   2010

Other aspects of the same approach, as well as a complementary bibliography, can be found in [*J. Jacod*, Calcul stochastique et problèmes de martingales. Lecture Notes in Mathematics. 714. Berlin-Heidelberg-New York: Springer-Verlag (1979; Zbl 414.60053)]. The martingale approach to diffusion processes consists in studying the following "martingale problem": Given functions $a : [0, \infty) \times \mathbb{R}^d \to S^d$ (the set of symmetric ≥ 0 $d \times d$-matrices) and $b : [0, \infty) \times \mathbb{R}^d \to \mathbb{R}^d$, define the operator:

$$L_u = \frac{1}{2} \sum_{i,j} a_{i,j}(u,x) \frac{\partial^2}{\partial x_i \partial x_j} + \sum_i b_i(u,x) \frac{\partial}{\partial x_i}$$

A solution to the corresponding martingale problem, starting from (s,x), is a probability measure P on the space of continuous paths $C(\mathbb{R}_+; \mathbb{R}^d)$, such that (i) $P(x(t) = x, 0 \le t \le s) = 1$; (ii) $f(x(t)) - \int_0^t L_u f(x(u)) \, du$ is a P-martingale for all $f \in C_0^\infty(\mathbb{R}^d)$.

The following questions are studied by the authors: (i) existence of a solution; (ii) uniqueness; (iii) additional properties, in the case one has both existence and uniqueness. The first third of the book is concerned with other ways of constructing a diffusion process, which provide in some cases a solution to the martingale problem: the method of Kolmogorov backward equation, and the method of Ito stochastic differential equation. Then, the martingale problem is studied, starting with existence. Uniqueness, which is harder to prove and requires the non-degeneracy of the diffusion coefficient a, is studied in detail. The Feller property of the process is proved under similar conditions. Also, the Cameron-Martin-Girsanov formula is established. A chapter is devoted to the results by Yamada and Watanabe an Ito's uniqueness, and its implication for the uniqueness of a solution to the martingale problem. The last chapter is concerned with the non-unique case, where a "good" selection can be made among the solutions. Whereas the theory is first developed in the case of bounded coefficients, one of the last chapters is devoted to its extension to the case of locally bounded coefficients. Basically, the results are the same provided the process does not explode. Conditions for explosion and non explosion are given.

Author profile:

Varadhan, Srinivasa R.S.

Spellings: Varadhan, S.R.S. [115] Varadhan, S. [7] Varadhan, Srinivasa R.S. [5] Varadhan, S. R. S. [4] Varadhan, Srinivasa [1] Varadhan, S.R.Srinivasa [1] Varadhan, S.R. Srinivasa [1] Varadhan, S. R. Srinivasa [1]
Author-Id: varadhan.srinivasa-r-s
Publications: 135 including 12 Book(s) and 73 Journal Article(s)

MSC 2010

109	60 · Probability theory and stochastic processes
19	82 · Statistical mechanics, structure of matter
17	35 · Partial differential equations (PDE)
4	00 · General mathematics
4	76 · Fluid mechanics

more ...

Journals

31	Communications on Pure and Applied Mathematics
7	Communications in Mathematical Physics
3	The Annals of Probability
2	Journal of Statistical Physics
2	Probability Theory and Related Fields

more ...

Co-Authors

24	Donsker, Monroe D.
17	Stroock, Daniel W.
9	Papanicolaou, George C.
8	Olla, Stefano
6	Landim, Claudio

more ...

Publication Years

The last-but-one chapter is concerned with limit theorems: convergence in law of diffusion processes to a diffusion process, and convergence of Markov chains to a diffusion process (result called invariance theorem). Here, the martingale formulation proves to be very powerful since martingale methods are very useful in proving limit theorems. Some of these are part of the preliminary material in probability, which forms Chapter 1. An appendix at the end of the book establishes some estimates from the theory of singular integrals. A good part of the theory presented in the book does need these results from analysis.

This book gives a very complete treatment of the theory of diffusion processes in $\mathbb{R}^d$. The various connections with partial differential equations are explained. The text itself is complemented with a great number of exercises, for which the main ideas of proof are often indicated. The book is very carefully written. Many introductions and comments explain the ideas, which might unless be hidden behind the often very intricate technicalities. The mathematical rigour is kept at a very high level. For instance, the construction of Ito's stochastic integral is given without neither completing the σ-algebras, nor skipping the difficulty about anticipative null-measure sets. This book constitutes a very complete – and for some results unique – reference text on diffusion processes. It should prove to be very useful to both theoretical and applied probabilists. For instance, the limit theorems are very useful in the theories of stochastic control and filtering. Though all the results needed from the theory of Markov processes, martingales, weak convergence of probability measures are proved in Chapter 1, the reader should have a good background in probability theory as well as in analysis, with some elementary functional analysis. I am sure that this book will be used both by researchers and teachers, and will become a commonly used reference book among scientists interested in the theory of diffusion processes.   *Etienne Pardoux (Marseille)*

Zbl 1103.60005
Stroock, Daniel W.; Varadhan, S. R. Srinivasa
Multidimensional diffusion processes.
Classics in Mathematics. Berlin: Springer (ISBN 3-540-28998-4/hbk). xii, 338 p. EUR 39.95/net; £ 30.50; sFr 73.00; $ 49.95 (2006).

[The first edition (1979) has been reviewed in Zbl 426.60069.]
The main purpose of this classic book is to elucidate the martingale approach to the theory of Markov processes and to propose a study of diffusion theory in $\mathbb{R}^d$.

The book is organized as follows. Chapter 1 provides an introduction to those parts of measure and probability theories which are considered most important for an understanding of the book. Basic tools are developed which are necessary for the construction of measures on functional spaces and introduces some notions and results which will play an important role in what follows. In Chapters 2 and 3 basic notions and results concerning Markov processes and parabolic partial differential equations are referred. A sketch of the procedure of studying diffusion theory via the backward equation is given which turns out to be one of the more powerful and successful approaches to the subject. In Chapter 4 Itô's theory of stochastic integration is developed. In Chapter 5 this theory is applied for studying the equation $dx(t) = \sigma(t, x(t))d\beta(t) + b(t, x(t))dt$ where $\beta(\cdot)$ is a d-dimensional Brownian motion. In Chapter 6 the study of diffusion theory is begun and a basic existence theorem for solutions to the martingale problem is proved. The relationship between the martingale problem and the (strong) Markov property, as well as the formula of Cameron, Martin and Girsanov are shown. Chapter 7 contains a proof of the general theorem about uniqueness for the solution of martingale problem. Chapter 8 expands the investigation of the relationships between Itô's approach and the martingale problem. The results of Chapter 9 are various L^p-estimates for the transition probability function of the considered multidimensional diffusion process. Chapter 10 extends the martingale problem approach to the case of unbounded coefficients. Standard conditions that can be used to test for explosion are given. In Chapter 11 the stability results for the Markov processes are studied. These results can be naturally divided into two categories: convergence of Markov chains to diffusions and convergence of diffusions to other diffusions. Chapter 12 takes up the question of what can be done in those circumstances when existence of solutions to a martingale problem can be proved but uniqueness cannot. It is also shown that every solution to a given martingale problem can, in some sense, be built out of those solutions which are part of a Markov family. In the Appendix some results from outside probability theory and concerning to the theory of singular integrals that are relied in Chapters 7 and 9 are proved in order to make the book self-contained.   *Pavel Gapeev (Berlin)*

Zbl 456.55001
Granas, Andrzej
Points fixes pour les applications compactes: Espaces de Lefschetz et la théorie de l'indice. Suivi d'une annexe par Kazimierz Geba.
Seminaire de Mathematiques Superieures, 68. Montreal: Les Presses de l'Universite de Montreal. 189 p. $ 10.00 (1980).

This monograph is an exposition of the fixed point theory in a general setting applicable to metric ANRs, infinite polyhedra, open subsets of convex sets in a locally convex vector space and other large classes of spaces, based on the works of Lefschetz, Leray, Schauder, Browder, of the author himself and of others.
Chapters 1,2 and 3 describe various types of spaces suitable for the fixed point theory (ANRs, neighbourhood extension spaces, spaces approachable by polyhedra) and maps such as compact, locally compact, eventually compact maps and maps of compact attraction. Embedding theorems for various classes of spaces are also given. Chapter 4 contains general facts about the fixed point property. In Chapter 5 and 6, generalized trace and the Lefschetz

Author profile:

Granas, Andrzej

Spellings: Granas, Andrzej [51] Granas, A. [40]
Author-Id: granas.andrzej
Publications: 91 including 6 Book(s) and 79 Journal Article(s)

MSC 2010

26	**47** · Operator theory
23	**54** · General topology
23	**55** · Algebraic topology
20	**34** · Ordinary differential equations (ODE)
14	**49** · Calculus of variations and optimal control; optimization

more ...

Journals

13	Comptes Rendus de l'Académie des Sciences. Série I
6	Journal de Mathématiques Pures et Appliquées. Neuvième Série
6	Topological Methods in Nonlinear Analysis
5	Journal of Mathematical Analysis and Applications
3	Fundamenta Mathematicae

more ...

Co-Authors

11	Geba, Kazimierz
8	Lee, John W.T.
7	Frigon, Mariène
7	Górniewicz, Lech
7	Guenther, Ronald B.

more ...

Publication Years

number are defined. A space X is a Lefschetz space for a given class of maps if to each map $f : X \to X$ in this class its Lefschetz number $\Lambda(f)$ can be assigned and if $\Lambda(f) \neq 0$ implies that f has a fixed point. In Chapter 7, Lefschetz spaces for various classes of maps (mentioned above) are studied and general theorems of Lefschetz type are proved. In Chapter 8, various versions of fixed point theorems are studied for maps of a restricted dimension, maps of pairs, approximative ANR and for semi-flows. Chapter 9 contains results due to Bowszyc, Hajek and Halpern on the existence of periodic points of a map $f : X \to X$. In Chapter 10, the fixed point index of a map (as developed by Dold) is extended to the class of metric ANRs and its basic properties are proved, including the Hopf-Lefschetz theorem: If $f : X \to X$ is a compact map, then $\Lambda(f)$ is equal to the fixed point index of f. A version of the Hopf-Lefschetz theorem for maps of compact attraction is also given. Chapter 11 is an exposition (without proofs) of the finite codimensional cohomology theory H^∞ as constructed by Gęba and further developed by Gęba and the author, on the category whose objects are closed arid bounded subsets of a normed vector space and whose maps are compact perturbations of the inclusion maps. Representation and Alexander duality theorems for this theory are stated and applications to questions of non-linear analysis are indicated. This part of the monograph is closed with historical comments and a bibliography list divided into several topics according to the organization of the book. An appendix, written by K. Gęba (in English), contains an application of the finite codimensional cohomology theory to bifurcation theory. If E is an infinite dimensional Banach space, let $GL_c(E)$ be the group of invertible operators in E which are compact (linear) perturbations of the identity. It is shown that $GL_c(E)$ has exactly two components. Next, a proof of a theorem due to Itze on the multiplicity of the characteristic values of a compact operator is given. The proof is based on a lemma involving the connecting homomorphism $H^\infty \to H^\infty$ of a suitable Mayer-Vietoris sequence. It is also shown that related results of Krasnosel'skii and Rabinowitz are consequences of Itze's theorem.

Reviewer's remark. The set U on p.176 and 186 should be assumed connected. There is also a confusing misprint an p.176, line 12: this line should probably read: $X = \partial U \cup \{$all bounded components of $E \times R \setminus U$ different from $U\}$. *Jan Jaworowski*

Zbl 474.43004
Marcus, Michael B.; Pisier, Gilles
Random Fourier series with applications to harmonic analysis.
Annals of Mathematics Studies, 101. Princeton, New Jersey: Princeton University Press and University of Tokyo Press. V, 150 p. hbk: $ 23.00; pbk: $ 9.25 (1981).

This work investigates the a.s. uniform convergence of random Fourier series on locally compact groups. In order to get a flavour of the results, we consider in this review the Abelian case. Let G be a locally compact Abelian group, Γ denote the characters of G, $K \subset G$ be a compact symmetric neighborhood of 0 and let $A \subset \Gamma$ be countable. Let $\{\xi_\gamma\}_{\gamma \in A}$ be a Rademacher sequence, $\{g_\gamma\}_{\gamma \in A}$ a sequence of independent $N(0, 1)$ random variables and $\{\xi_\gamma\}$ complex valued random variables satisfying $\sup_{\gamma \in A} E|\xi_\gamma|^2 < \infty$, $\inf_{\gamma \in A} E|\xi_\gamma| > 0$. For a sequence of complex numbers $\{\delta_\gamma\}$ in $\ell^2(A)$, let $\gamma \in A$

$$Z(x) = \sum_{\gamma \in A} \delta_\gamma \varepsilon_\gamma \xi_\gamma(x), \quad x \in K. \tag{$*$}$$

The a.s. uniform convergence of this series is controlled by the pseudo metric (on $K \oplus K$), $\sigma(x - y) = (\sum_{\gamma \in A} |\delta_\gamma|^2 |\gamma(x) - \gamma(y)|^2)^{1/2}$. Consider σ as a function on K, let $\tilde\sigma$ denote the nondecreasing rearrangement of σ and let

$$I(\sigma) = \int_0^{\mu_n} \frac{\tilde\sigma(s)}{s \left(\log \frac{4\mu}{s}\right)^{1/2}} \, ds,$$

where μ is the Haar measure an G and $\mu_n = \mu(\bigoplus_n K)$. Then, if $I(\sigma) < \infty$, the series $(*)$ converges uniformly a.s., and

Author profile:

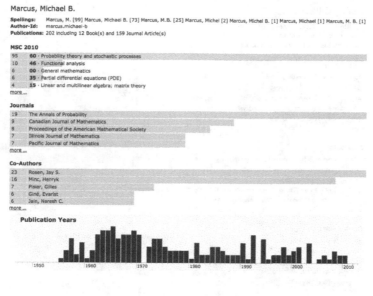

Marcus, Michael B.

Spellings: Marcus, M. [99] Marcus, Michael B. [73] Marcus, M.B. [25] Marcus, Michel [2] Marcus, Michel B. [1] Marcus, Michael [1] Marcus, M. B. [1]
Author-Id: marcus.michael-b
Publications: 202 including 12 Book(s) and 159 Journal Article(s)

$$(E\|Z\|_\infty^2)^{1/2} \leq C \left(\sup_{\gamma \in A} E|\xi_\gamma|^2\right)^{1/2} \left(\left\|\{\delta_\gamma\}\right\|_{\ell^2(A)} + I(\sigma)\right)$$

where C is a constant independent of $\{\delta_\gamma\}$ and $\{\gamma \mid \gamma \in A\}$. There is also a converse to this result. This material is developed in Chapter 3. The next chapter deals with (*) when viewed as a $C(K)$ random variable. It is shown that, if $I(\sigma) < \infty$, then Z satisfies the central limit theorem and the law of the iterated logarithm. Chapter 5 deals with non-commutative versions of the above results. Chapter 6 contains several applications of random Fourier series to harmonic analysis. Let G be a compact group, and let $C_{a.s.}(G) = \{f \in L_2(G) : \text{random Fourier series of } f \text{ is a. s. continuous}\}$. One of the main results is that

$$C_{a.s.}(G) = \left\{ f = \sum_{n=1}^\infty h_n * k_n, \, h_n \in L_2(G), \, k_n \in D\sqrt{\log L(G)}, \, \sum_{n=1}^\infty \|h_n\|_2 \|k_n\| < \infty \right\}.$$

In the last chapter, the authors return to classical grounds and provide new proofs of some of the classical results of Paley-Zygmund-Salem-Kahane which motivated their work. To conclude, let me quote the authors: "We hope that these results will be of interest to probabilists who can view random Fourier series as interesting and useful examples of stochastic processes – intimately related to stationary Gaussian processes – and to harmonic analysts who may find these results and methods useful in their own work."

M. Milman

146

Author profile:

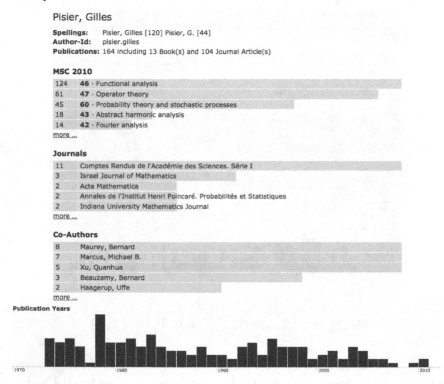

Zbl 483.10001
Hua, Loo Keng
Introduction to number theory. Translated from the Chinese by Peter Shiu.
Berlin-Heidelberg-New York: Springer-Verlag. XVIII, 572 p., 14 figs. DM
96.00; $ 42.70 (1982).

Das chinesische Original der vorliegenden Übersetzung stammt aus dem Jahr 1957, wurde 1964 unverändert nachgedruckt und in Zbl 221.10002 referiert. Wie der Autor bekennt, ist das Buch weniger als Begleitlektüre etwa zu einer systematischen Einführungsvorlesung in die Zahlentheorie gedacht als vielmehr als Kompendium, welches den an zahlentheoretischen Fragestellungen Interessierten möglichst rasch und auf breiter Front bis an die (1957) neueste Forschung auf diesem Gebiet heranführen sollte. Damit auch die nun präsentierte englische Übersetzung diesem hohen Anspruch möglichst gerecht wird, hat Wang Yuan den meisten der 20 Kapitel, deren Überschriften dem o.g. Referat entnommen werden können, kurze Anhänge beigefügt, die die Entwicklung in den 25 Jahren seit dem Original berücksichtigen. Dieses wertvolle Buch stellt eine bedeutende Ergänzung der Literatur zur Zahlentheorie dar. Es kann kapitelweise zum zügigen Vertrautwerden mit dem Gegenstand benutzt werden, bringt andererseits aber Ergebnisse vor allem der analytischen Theorie wie sie sonst in einer so breit an-

Author profile:

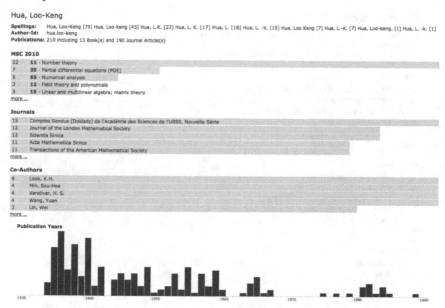

gelegten "Einführung" kaum zu finden sind. Hierfür seien etwa als Beispiele genannt: Vinogradovs Methode zur Abschätzung von Summen über den gebrochenen Anteil gewisser reeller Zahlen und Anwendung dieser Methode auf Gitterpunktprobleme; das Resultat von Erdös und Fuchs; elementarer Beweis nach Erdös und Selberg für den Primzahlsatz und dessen analytischer Beweis via Ikeharas Taubersatz; Siegels Satz über die Klassenzahl imaginärquadratischer Zahlkörper; Gelfonds Lösung des siebten Hilbertschen Problems sowie Selbergs Siebmethode und der Satz von Goldbach-Schnirelman. Da und dort eingestreut in den Text finden sich Übungsaufgaben für den Leser, aber auch einige nützliche Tabellen wie z.B. eine solche für die kleinsten Primitivwurzeln modulo der Primzahlen unterhalb 5000.

Versucht man dieses schöne und überaus umfassende Zahlentheoriebuch in die existierende Literatur einzuordnen, so wird einem sein Charakter am ehesten an den Klassiker von *G. H. Hardy* und *E. M. Wright* [An introduction to the theory of numbers. 5th ed. Oxford etc.: Oxford at the Clarendon Press (1979; Zbl 423.10001)] erinnern, wenngleich Huas Buch inhaltlich weit hierüber hinausgeht. *Peter Bundschuh (Köln)*

Zbl 528.57009
Thurston, William P.
Three dimensional manifolds, Kleinian groups and hyperbolic geometry.
The mathematical heritage of Henri Poincaré, Proc. Symp. Pure Math. 39,
Part 1, Bloomington/Indiana 1980, 87-111 (1983).

In the last few years the face of three dimensional topology has changed completely, due mainly to deep and fascinating discoveries of the author

148

Author profile:

Thurston, William P.

Spellings: Thurston, William P. [37] Thurston, William [14] Thurston, W.P. [10] Thurston, W. [8]
Author-Id: thurston.william-p
Publications: 69 including 3 Book(s) and 52 Journal Article(s)

MSC 2010

43	**57** · Manifolds and cell complexes
13	**53** · Differential geometry
9	**20** · Group theory and generalizations
8	**55** · Algebraic topology
7	**37** · Dynamical systems and ergodic theory

more ...

Journals

5	Inventiones Mathematicae
5	Topology
4	Annals of Mathematics. Second Series
3	Bulletin (New Series) of the American Mathematical Society
3	Geometry & Topology

more ...

Co-Authors

5	Hass, Joel
3	Agol, Ian
3	Cannon, James W.
3	Dunfield, Nathan M.
3	Miller, Gary Lee

more ...

Publication Years

who had revealed for topologists an entirely new world of beautiful geometrical phenomena. The field of three dimensional topology had remained relatively isolated for a long time. Now many deep connections of three dimensional topology with hyperbolic geometry, theory of Kleinian groups and theory of Riemannian surfaces come to light. Due to the author's work, our understanding of 3-manifolds is much bester now than a few years ago. Indeed, the author proposed a very clear picture of 3-manifolds, which is succinctly formulated as his "Geometrization Conjecture" (Conjecture 1.1 in the reviewed paper): The interior of every compact 3-manifold has a canonical decomposition into pieces which have geometric structures. This conjecture generalizes the Poincaré conjecture. Its advantage over the Poincaré conjecture arises from the fact that it applies to all 3-manifolds, so there are many particular cases to check this conjecture. The author proved this conjecture in many important cases, the most notable one is the case of Haken manifolds (for example, exteriors of all knots in S^3 are Haken manifolds). There are eight types of geometric structures in question. The most interesting, important and complicated structure is the hyperbolic one (a hyperbolic manifold is a Riemannian manifold of constant negative curvature).

The reviewed paper is a survey of the author's work in this field. The paper is intended for a wide audience but extremely interesting for specialists in

three dimensional topology or Kleinian groups, too. The paper is beautifully illustrated with excellent pictures by G. Francis. The complete proofs
of the author's results are still unpublished, and this paper gives a good
introduction to this circle of ideas.

Firstly, the Geometrization Conjecture is precisely formulated and discussed.
After this many cases in which the Conjecture is proved are discussed. A
new exciting result is formulated in a footnote added in proof: the conjecture is true for all prime 3-manifolds with a symmetry having fixed point
set of dimension ≥ 1. The next section is devoted to applications. Among
them is the proof of the Smith conjecture an fixed point sets of periodic diffeomorphisms of S^3 (the proof of the Smith conjecture is the result of joint
efforts of several mathematicians). Further, the eight 3-dimensional geome-
tries are briefly described. The next topic is the geometry of Kleinian groups
and its connection with the geometry and topology of 3-manifolds and with
the structure of surface diffeomorphisms. The final section of the paper is
devoted to 24 questions and projects concerning 3-manifolds and Kleinian
groups. Just before this the hope is expressed that "perhaps by the year 2000
our understanding of 3-manifolds and Kleinian groups will be solid, and
the phenomena we now expect will be proved." Indeed, these questions
constitute a program, the realization of which will lead to the realization of
this hope.   *Nikolai V. Ivanov*

Zbl 588.14026
Faltings, G.
Finiteness theorems for abelian varieties over number fields.
Invent. Math. 73, 349-366 (1983); Erratum ibid. 75, 381 (1984).

Der Autor hat in dieser Arbeit seinen vielbeachteten und gefeierten Beweis
der Mordellschen Vermutung niedergelegt: daß jede glatte projektive algebraische Kurve X vom Geschlecht $g \geq 2$ über einem Zahlkörper K nur endlich
viele K-rationale Punkte $X(K)$ hat (Satz 7). Dank einer früheren Arbeit von
A. N. Parshin [Math. USSR, Izv. 2 (1968), 1145–1170 (1970); Übersetzung von
Izv. Akad. Nauk SSSR, Ser. Mat. 32, 1191–1219 (1968; Zbl 181.23902)] genügte
es dafür, die sogenannte Shafarevich-Vermutung zu beweisen:
Es gibt nur endlich viele (bis auf Isomorphie) glatte projektive algebraische
Kurven X über K mit vorgegebenem Geschlecht g und guter Reduktion
außerhalb einer vorgegebenen endlichen Menge S von Primstellen von K.
Durch Übergang zu den Jacobischen folgt diese Aussage mit Hilfe einer
geeigneten Fassung des Satzes von Torelli aus der Aussage (Satz 6):
Es gibt nur endlich viele Isomorphieklassen abelscher Varietäten mit Po-
larisierung von vorgegebenem Grad $d > 0$ über K, die außerhalb S gute
Reduktion haben.
Dies beweist der Autor und, auf dem Wege dahin, auch die wichtige, vorher
nur in wenigen Spezialfällen bewiesene, Tate-Vermutung für Endomorphis-
men abelscher Varietäten A über K:

150

Author profile:

Faltings, Gerd

Spellings: Faltings, Gerd [68] Faltings, G. [7]
Author-Id: faltings.gerd
Publications: 75 including 7 Book(s) and 44 Journal Article(s)

MSC 2010

69	**14** · Algebraic geometry
26	**11** · Number theory
7	**13** · Commutative algebra
6	**32** · Functions of several complex variables and analytic spaces
2	**00** · General mathematics

more ...

Journals

6	Inventiones Mathematicae
6	Journal of Algebraic Geometry
5	Journal für die Reine und Angewandte Mathematik
4	Mathematische Annalen
3	Annals of Mathematics. Second Series

more ...

Co-Authors

5	Wüstholz, Gisbert
1	Chai, Ching-Li
1	de Jong, Johan P.
1	Jordan, Bruce W.
1	Pink, Richard

Publication Years

Sei $\pi = \mathrm{Gal}(\bar{K}/K)$ die absolute Galoisgruppe von K, ℓ eine Primzahl. Der inverse Limes der ℓ^n-Teilungspunkte von A, $T_\ell(A) = \lim_{\leftarrow n} A[\ell^n](\bar{K})$, ist ein freier $\mathbb{Z}_\ell$-Modul mit π-Aktion. Dann gilt:

(Satz 3) Die Darstellung von π auf $T_\ell(A) \otimes_{\mathbb{Z}_\ell} \mathbb{Q}_\ell$ ist halbeinfach.

(Satz 4) Die natürliche Abbildung $\mathrm{End}_K(A) \otimes \mathbb{Z}_\ell \to \mathrm{End}_\pi(T_\ell(A))$ ist ein Isomorphismus.

Diese Aussagen wurden von *J. Tate* [Invent. 2, 134–144 (1966; Zbl 147.20303)] für abelsche Varietäten über endlichen Körpern bewiesen, und Tate's Ansatz wurde von *Yu. G. Zarkhin* [insbesondere in Math. USSR, Izv. 9 (1975), 255–260 (1976); Übersetzung von Izv. Akad. Nauk SSSR, Ser. Mat. 39, 272–277 (1975; Zbl 345.14014)] für den Funktionenkörperfall ausgebaut. Mit denselben Methoden (Zarkhin, loc. cit.; nicht die von ihm zitierten Arbeiten Zarkhins) reduziert der Autor hier den Zahlkörperfall auf Eigenschaften der Höhe einer abelschen Varietät über K, $h(A)$.

Dieser Höhenbegriff ist die entscheidende begriffliche Neuheit der Arbeit. Zwar ist $h(A)$ im wesentlichen nur ein – durch die Arakelovtheorie motiviertes – Maß der Volumina von $A(\overline{K_v})$ bzgl. eines Néron-Differentials auf A, wo K_v die archimedischen Komplettierungen von K durchläuft. Der Autor beweist aber für $h(A)$ die typische diophantische Eigenschaft einer Höhe auf dem Modulraum A_g prinzipal polarisierter abelscher Varietäten der Dimension g:

Gegeben c, es existieren nur endlich viele Isomorphieklassen prinzipal polarisierter abelscher Varietäten A der Dimension g über K, mit semistabiler Reduktion, so daß $h(A) \leq c$ ist (Satz 1).

Der Beweis dieses Satzes ist der technisch aufwendigste Teil der Arbeit (§§ 2,3): A_g is nicht leicht zu handhaben (an entscheidender Stelle greift der Autor noch auf einen Vergleich mit dem Modulschema stabiler Kurven vom Geschlecht g zurück), und h hat – wenn auch nur logarithmische – Singularitäten längs des Randes der Kompaktifizierung von A_g.

Für hilfreiche Varianten dieses Satzes 1 und seines Beweises, wie auch für einen laufenden Kommentar aller Teile der Arbeit, empfiehlt sich sehr der Vergleich mit den verschiedenen Ausarbeitungen von des Autors Argumenten, die inzwischen erschienen sind: *P. Deligne*, Sémin. Bourbaki, 36e année, 1983/84, Exp. No. 616, Astérisque 121/122, 25–41 (1985; Zbl 591.14026) and *L. Szpiro*, Sémin. Bourbaki, 36e année, 1983/84, Exp. No. 619, Astérisque 121/122, 83–103 (1985; Zbl 591.14027); *G. Faltings, G. Wüstholz* et. al., "Rational points", Semin. Bonn/Wuppertal 1983/84 (1984; Zbl 588.14027); *L. Szpiro*, "Séminaire sur les pinceaux arithmétiques: La conjecture de Mordell", Astérisque 127 (1985; Zbl 588.14028); *G. Cornell* und *J. H. Silverman*, Arithmetic Geometry, Proc. Conf. Storrs 1984 (Springer Verlag 1986; Zbl 596.00007) (das letztgenannte Buch enthält eine getreue englische Übersetzung des hier referierten Artikels, s. Zbl 602.14044). Vgl. auch *Yu. G. Zarkhin*, Invent. Math. 79, 309–321 (1985; Zbl 557.14024).

Die weiteren Schaltstellen im Beweisgang sind wie folgt: Um Sätze 3 und 4, gemäß Tates Ansatz, auf Satz 1 zu reduzieren, wird das Verhalten der Höhe $h(A)$ berechnet, wenn A durch die Stufen einer über K definierten ℓ-divisiblen Untergruppe (G_n) von $(A[\ell^n])$ geteilt wird: Die Folge dieser Höhen wird stationär (Satz 2/Erratum). Hier wird Tates Strukturtheorie ℓ-divisibler Gruppen herangezogen.

Um zu Satz 6 zu gelangen, wird zunächst die Endlichkeit der Anzahl der Isogenieklassen abelscher Varietäten der Dimension g über K mit guter Reduktion außerhalb S gezeigt (Satz 5): Nach Satz 3/4 ist A bis auf K-Isogenie bestimmt durch die Spuren fast aller Frobenius-Elemente F_v ($v \nmid \ell$) auf $T_\ell(A)$, oder durch fast alle Euler Faktoren $L_v(A, s)$ der Hasse-Weil L-Funktion von A über K. Nach Weil's Satz über die Eigenwerte der F_v gibt es nur endlich viele Möglichkeiten für jedes $L_v(A, s)$. Der Autor zeigt nun mit einem überraschenden, kurzen Argument, wie Chebotarev's Dichtigkeitssatz A schon durch die L_v, für v in einer geeigneten endlichen Menge von Stellen von K, bis auf Isogenie bestimmt. Das erledigt Satz 5.

Für Satz 6 bleibt zu zeigen, daß jede Isogenieklasse in nur endlich viele Isomorphieklassen zerfällt. Unter Benutzung von Satz 3/4 sowie Satz 1 reduziert sich dies bald auf eine Höhenberechnung, die der im Beweis von

152

Satz 2 analog ist, allerdings Raynauds Theorie [*M. Raynaud*, Bull. Soc. Math. Fr. 102 (1974), 241–280 (1975; Zbl 325.14020)] benutzt.
Vgl. auch das Referat über den Übersichtsartikel des Autors in Jahresber. Dtsch. Math.-Ver. 86, 1–13 (1984; Zbl 586.14012).   *Norbert Schappacher*

Zbl 568.20001
Conway, J.H.; Curtis, R.T.; Norton, S.P.; Parker, R.A.; Wilson, R.A.
Atlas of finite groups. Maximal subgroups and ordinary characters for simple groups. With computer assistance from J. G. Thackray.
Oxford: Clarendon Press. XXXIII, 252 p. £35.00 (1985).

Author profile:

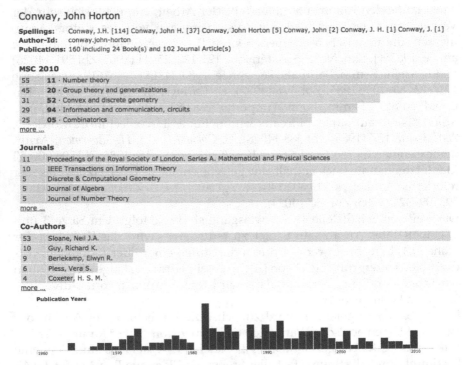

Conway, John Horton

Spellings: Conway, J.H. [114] Conway, John H. [37] Conway, John Horton [5] Conway, John [2] Conway, J. H. [1] Conway, J. [1]
Author-Id: conway.john-horton
Publications: 160 including 24 Book(s) and 102 Journal Article(s)

MSC 2010

55	11 · Number theory
45	20 · Group theory and generalizations
31	52 · Convex and discrete geometry
29	94 · Information and communication, circuits
25	05 · Combinatorics

more ...

Journals

11	Proceedings of the Royal Society of London. Series A. Mathematical and Physical Sciences
10	IEEE Transactions on Information Theory
5	Discrete & Computational Geometry
5	Journal of Algebra
5	Journal of Number Theory

more ...

Co-Authors

53	Sloane, Neil J.A.
10	Guy, Richard K.
9	Berlekamp, Elwyn R.
6	Pless, Vera S.
4	Coxeter, H. S. M.

more ...

Publication Years

This book (which in the remainder of this report we will just call the ATLAS) is in its main part a table collection on finite simple groups. Its purpose is to convey detailed information about specific groups. The following informations about a simple group G are provided:

(1) The order of G, the structure of its Schur multiplier, and its outer automorphism group.

(2) Maximal subgroups of G respectively X ($G \leq X \leq Aut(G)$).

(3) Various constructions of G respectively of groups Y, where $Z(Y) \leq Y^{(\infty)}$, and $Y/Z(Y)$ is isomorphic to a group X ($G \leq X \leq Aut(G)$).

(4) Presentations of G or Y (Y as in (3)).

(5) Character tables of all groups Y (Y as in (3)).

Author profile:

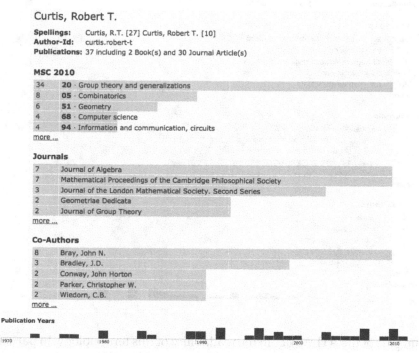

Curtis, Robert T.

Spellings: Curtis, R.T. [27] Curtis, Robert T. [10]
Author-Id: curtis.robert-t
Publications: 37 including 2 Book(s) and 30 Journal Article(s)

MSC 2010

34	**20** · Group theory and generalizations
8	**05** · Combinatorics
6	**51** · Geometry
4	**68** · Computer science
4	**94** · Information and communication, circuits
more ...	

Journals

7	Journal of Algebra
7	Mathematical Proceedings of the Cambridge Philosophical Society
3	Journal of the London Mathematical Society. Second Series
2	Geometriae Dedicata
2	Journal of Group Theory
more ...	

Co-Authors

8	Bray, John N.
3	Bradley, J.D.
2	Conway, John Horton
2	Parker, Christopher W.
2	Wiedorn, C.B.
more ...	

Publication Years

The only mayor information the ATLAS does not provide are modular characters. The list of finite simple groups in the ATLAS has about 100 members and includes all 26 sporadic simple groups. The alternating groups occur up to degree 13. From the classical linear groups the "first members" of a family are listed; this means that the Lie-rank is less or equal to 5 and the field is small (i.e., if the Lie-rank is ≥ 3, then the field is GF(2) or GF(3)). The exceptional Chevalley groups are only included for small fields (i.e., GF(2) if the Lie-rank is ≥ 3) too. Thus with the exception of small rank Chevalley groups all simple groups of order less than 10^{25} are listed.

There are obvious difficulties to present such an impressive amount of material economically. The authors of the ATLAS were forced to find a compromise between the necessity of presenting the material completely while keeping the size of the book to a minimum, and to make the tables understandable for a wider readership. Thus in the introduction to the ATLAS there is first a small general description of the finite simple groups and the way they were originally constructed. In the second part of the introduction the authors introduce some terminology of their own to give the desired economic presentation. The critical point in understanding the tables in the

Author profile:

Norton, Simon P.

Spellings: Norton, S.P. [14] Norton, Simon P. [8] Norton, Simon [5] Norton, S. [2]
Author-Id: norton.simon-p
Publications: 29 including 1 Book(s) and 13 Journal Article(s)

MSC 2010

26	**20** · Group theory and generalizations
8	**11** · Number theory
3	**17** · Nonassociative rings and algebras
2	**05** · Combinatorics
1	**03** · Mathematical logic

more ...

Journals

3	Communications in Algebra
1	Bulletin de la Société Mathématique de Belgique. Série A
1	Bulletin of the London Mathematical Society
1	Canadian Journal of Mathematics
1	Discrete Mathematics

more ...

Co-Authors

4	Wilson, Robert A.
3	Conway, John Horton
2	Glauberman, George
1	Cummins, C. J.
1	Curtis, Robert T.

more ...

Publication Years

1970   1980   1990   2000   2010

main part of the ATLAS lies in the readability of this terminology. In particular the notation for character tables of groups Y, where Y has both nontrivial Schur multiplier and outer automorphism group, becomes quite intricate. To understand these tables fully will not be always immediate! However there are some character tables in this introduction, which are presented in the usual way and secondly in the new "stenograph type" notation.

In the main part of the ATLAS for each of the listed groups there is an extra section. First a group G is listed by all different names under which it occurs thus making exceptional isomorphisms visible (for instance A_6 is also listed as $L_2(9)$, $U_2(9)$, $S_2(9)$, $O_3(9)$, $O_4^-(3)$, $S_4(2)'$, $O_5(2)'$). Secondly methods of constructing G (or a covering group) are given. This is done in a very brief way. The reader must usually put in some extra work to fill in the details. Then a presentation with generators and relations is given (this presentation is one which is easy to compute with but normally not one with a minimum number of generators or relations). The maximal subgroups (normally all) are given not only in a structural description but also in the role they play in various constructions. The largest part for a group G will usually be the character tables for the groups Y with $Y/Z(Y) \simeq X$ (where $G \leq X \leq Aut(G)$). These tables contain even subtle details: the Schur indicator (which tells us whether a character is nonreal, or real but comes from a nonreal

Author profile:

Parker, Richard A.

Spellings: Parker, R.A. [8] Parker, Richard A. [4] Parker, R. [3] Parker, Richard [2]
Author-Id: parker.richard-a
Publications: 17 including 3 Book(s) and 12 Journal Article(s)

MSC 2010

8	**20** · Group theory and generalizations
2	**65** · Numerical analysis
2	**68** · Computer science
2	**92** · Applications of mathematics to Biology and other natural sciences
1	**01** · History; biography

more ...

Journals

2	International Journal for Numerical Methods in Engineering
2	Journal of Algebra
1	Information Processing Letters
1	International Journal of Systems Science
1	Journal of Mathematical Physics

more ...

Co-Authors

4	Wilson, Robert A.
2	Benson, David John
2	Conway, John Horton
1	Bose, Sanjay K.
1	Camassa, Roberto

more ...

Publication Years

representation, or a character of a real representation) or the conjugacy classes in which the powers of a given representative of a conjugacy class fall etc. The ATLAS closes with some appendices among which one finds a comprehensive bibliography on finite, sporadic simple groups.

One should emphasize that the ATLAS gives information about individual groups. Usually it should be difficult to extrapolate from the numerical data the ATLAS provides to a whole infinite series of finite simple groups. The ATLAS gives a kind of information, which in published papers on finite simple groups is hard to get, but which is so useful in a concrete situation. The ATLAS should provide for a large readership of mathematicians the quickest and most efficient way to get that kind of information. It also might be just fun (like for the reviewer) to run at random through the ATLAS and find new surprising details about groups which seemed familiar. In conclusion: This ATLAS will be of greatest importance for any mathematician who faces concrete problems in finite simple groups. *U. Dempwolff*

156

Author profile:

Wilson, Robert A.

Spellings: Wilson, Robert A. [82] Wilson, R. [76] Wilson, R.A. [20]
Author-Id: wilson.robert-a
Publications: 178 including 4 Book(s) and 163 Journal Article(s)

MSC 2010

97	**20** · Group theory and generalizations
5	**30** · Functions of a complex variable
4	**51** · Geometry
4	**68** · Computer science
3	**57** · Manifolds and cell complexes

more ...

Journals

22	Journal of Algebra
16	Journal of the London Mathematical Society
15	Proceedings of the London Mathematical Society. Second Series
11	Proceedings of the Edinburgh Mathematical Society. Series II
9	Mathematical Proceedings of the Cambridge Philosophical Society

more ...

Co-Authors

11	Suleiman, Ibrahim A.I.
8	Kleidman, Peter B.
6	Bray, John N.
6	Macintyre, Archibald J.
5	Holmes, Petra E.

more ...

Publication Years

Zbl 632.43001
Schempp, W.
Harmonic analysis on the Heisenberg nilpotent Lie group, with applications to signal theory.
Pitman Research Notes in Mathematics Series, 147. Harlow, Essex, England: Longman Scientific & Technical. Copubl. in the United States with John Wiley & Sons, Inc., New York. VIII, 199 p. £15.00 (1986).

This book presents the theory of harmonic analysis on nilpotent groups with special emphasis on the real Heisenberg group $\tilde{A}(\mathbb{R})$. In particular the unitary dual of $\tilde{A}(\mathbb{R})$ is determined by an application of the unitary inducing procedure, called the Mackey machinery, and the Kirillov orbit picture. Furthermore the author establishes a connection to signal theory: the coefficients of a suitable representation of $\tilde{A}(\mathbb{R})$ are related to the radar ambiguity functions. Among the results, obtained in this way, is the solution of the synthesis problem.

Chapter I contains the definitions and basic facts of representation theory for locally compact topological groups. In chapter II various inducing procedures for representations are given, in particular the Mackey machinery is introduced. Square integrable group representations are studied in chap-

Author profile:

Schempp, Walter Johannes

Spellings: Schempp, Walter [169] Schempp, W. [8] Schempp, Walter Johannes [2] Schempp, Walter J. [1]
Author-Id: schempp.walter-johannes
Publications: 180 including 12 Book(s) and 98 Journal Article(s)

MSC 2010

62	41 · Approximation and expansions
60	43 · Abstract harmonic analysis
42	22 · Topological groups, Lie groups
32	94 · Information and communication, circuits
30	42 · Fourier analysis

more ...

Journals

17	Comptes Rendus Mathématiques de l'Académie des Sciences
13	Mathematische Zeitschrift
10	Results in Mathematics
9	Journal of Approximation Theory
3	Acta Applicandae Mathematicae

more ...

Co-Authors

17	Dreseler, Bernd
14	Delvos, Franz-Jürgen
13	Binz, Ernst
4	Zeller, Karl
3	Pods, Sonja

more ...

Publication Years

ter III, leading to the orthogonality relations for coefficients of irreducible representations when integrated over G modulo C, the centre. Chapter IV is devoted to nilpotent topological groups. Here every irreducible representation can be constructed by inducing 1-dimensional representations of suitable subgroups.

The real Heisenberg group $\tilde{A}(\mathbb{R})$ is an example of a locally compact, two-step nilpotent, topological group having a non-trivial centre. In chapter V various presentations of this group are given and its unitary dual is explicitly determined. The irreducible unitary representations occur as actions on $L^2(\mathbb{R})$. The next chapter introduces the coadjoint orbit picture due to Kirillov, which relates the irreducible representations of G to the orbits of the action of G on $\mathfrak{g}^*$, the dual vector space of the Lie algebra Lie(G). Applying this method to $\tilde{A}(\mathbb{R})$ gives again the unitary dual along with an explicit calculation of the corresponding eigenfunctions.

In the last chapter the author exploits the connections between representation theory and signal theory. It is observed that the radar cross ambiguity function $H(f, g; x, y) = \int f(t + 1/2x) * \bar{g}(t - 1/2x) * e^{2\pi i y t} dt$ occurs as the coefficient function $c_{f,g}(x, y, 0)$ of the $\tilde{A}(\mathbb{R})$-representation on $L^2(\mathbb{R})$, $n = 3$. The author then studies the fiber bundle of G over its centre C, which gives a solution to the synthesis probem: any function $F \in \mathcal{S}(\mathbb{R} + \mathbb{R})$ occurs as a radar cross ambiguity function for some $f, g \in S(\mathbb{R})$. The functions on the diagonal, i.e. the radar auto-ambiguity functions H(f,f;x,y), are uniquely characterized to be minimal and of positive type and the auto- ambiguity functions, which are $SO(2, \mathbb{R})$-invariant, are multiples of the Laguerre functions.   *P. Maaß*

Zbl 624.52004
Kalai, Gil
Rigidity and the lower bound theorem. I.
Invent. Math. 88, 125-151 (1987).

Author profile:

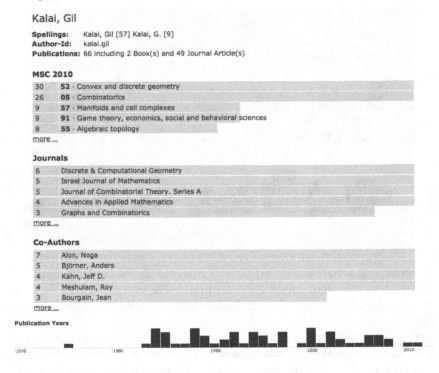

Kalai, Gil

Spellings: Kalai, Gil [57] Kalai, G. [9]
Author-Id: kalai.gil
Publications: 66 including 2 Book(s) and 49 Journal Article(s)

MSC 2010

30	**52** · Convex and discrete geometry
26	**05** · Combinatorics
9	**57** · Manifolds and cell complexes
9	**91** · Game theory, economics, social and behavioral sciences
8	**55** · Algebraic topology

more ...

Journals

6	Discrete & Computational Geometry
5	Israel Journal of Mathematics
5	Journal of Combinatorial Theory. Series A
4	Advances in Applied Mathematics
3	Graphs and Combinatorics

more ...

Co-Authors

7	Alon, Noga
5	Björner, Anders
4	Kahn, Jeff D.
4	Meshulam, Roy
3	Bourgain, Jean

more ...

Publication Years

Barnette's lower bound theorem (LBT) says that for every triangulated $(d-1)$-manifold with n vertices the f-vector satisfies the inequality $f_k \geq \phi_k(n,d)$ where $\phi_k(n,d) := \binom{d}{k} - \binom{d+1}{k+1}k$ for $1 \leq k \leq d - 2$ and $\phi_{d-1}(n,d) := (d-1)n - (d + 1)(d-2)$. It has been conjectured that equality $f_k = \phi_k(n,d)$ for some k implies that the manifold is a sphere combinatorially equivalent to the boundary complex of a stacked d-polytope. The main result of the present paper is another proof of the LBT and a proof of this conjecture about the case of equality. As expressed by the title the author uses a rigidity result saying that almost all embeddings of the 1-skeleton of such a manifold into $\mathbb{R}^d$ are rigid (this is called "generically d-rigid"). The author remarks that the basic relation between the LBT and rigidity has been observed independently by M. Gromov.

Among other results the author discusses manifolds with stacked links, one result saying that every $(d-1)$-manifold $(d \geq 5)$ which has only stacked vertex links, must belong to the class $\mathcal{H}^d$ defined by D. Walkup [Acta Math. 125, 75-107 (1970; Zbl 204.56301)], in particular it cannot be simply connected. Related results about manifolds with boundary, pseudomanifolds and polyhedral manifolds are also included.

At the end various conjectures are given. One of them relates the number $\gamma(M) := \min\{f_1 - \phi_1(n,d)\}$ with the first Betti number of M, another one concerns a version of the generalized lower bound conjecture (GLBC) involving the h-vector and the Betti numbers. This remarkable paper is announced to continue with parts II and III.   *W. Kühnel*

Zbl 648.94011
van Lint, Jacobus H.
Introduction to coding theory and algebraic geometry. I: Coding theory.
Coding theory and algebraic geometry, Lect. Semin., Düsseldorf/FRG 1987, DMV Semin. 12, 9-33 (1988).

[For the entire collection see Zbl 639.00048.]

An exciting new development in algebraic coding theory is concerned with the application of methods from algebraic geometry to the construction of linear block codes. The principal achievement so far of this line of research is the construction of families of linear block codes that meet or even go beyond the classical Gilbert-Varshamov bound. In 1987 the Deutsche Mathematiker-Vereinigung organized a seminar on this topic which featured two series of lectures, one on the coding theory background and the other on the algebraic geometry background. The paper under review is based on the first series of lectures. It contains a masterful exposition of the theory of linear block codes which requires only minimal algebraic prerequisites. Besides basic concepts, the article covers BCH codes, classical Goppa codes, bounds on codes such as the Gilbert-Varshamov bound and the Plotkin bound, self-dual codes, and some examples of codes obtained from algebraic curves. The elegant, succinct style allows the presentation of a lot of material in a relatively short space.   *H. Niederreiter.*

160

Author profile:

van Lint, Jacobus Hendricus

Spellings: van Lint, J.H. [112] van Lint, Jacobus H. [12] van Lint, Jack [3] van Lint, Jack H. [2] van Lint, J.H. jun. [2] Van Lint, J.H. [2] [1]
Author-Id: van-lint.jacobus-hendricus
Publications: 136 including 25 Book(s) and 87 Journal Article(s)

MSC 2010

66	**94** · Information and communication, circuits
61	**05** · Combinatorics
13	**11** · Number theory
8	**00** · General mathematics
8	**51** · Geometry

more ...

Journals

12	Journal of Combinatorial Theory. Series A
10	IEEE Transactions on Information Theory
10	Nederlandse Akademie van Wetenschappen. Proceedings. Series A. Indagationes Mathematicae
4	Designs, Codes and Cryptography
3	Discrete Mathematics

more ...

Co-Authors

6	Cameron, Peter J.
5	Hollmann, Henk D.L.
5	Wilson, Richard M.
4	de Bruijn, Nicolaas Govert
4	Hall, Marshall jun.

more ...

Publication Years

Zbl 717.53062
Bourguignon, Jean-Pierre
A mathematician's visit to Kaluza-Klein theory
Rend. Semin. Mat., Torino, Fasc. Spec., 143-163 (1989).

Kaluza-Klein theory began as an attempt to unify gravitation (represented by a metric) and electromagnetism (described by a one form, α) in the early days of general relativity. This ingenious approach involved extending the space-time model from a four-manifold M, to a five-dimensional one, $\tilde{M}$. The four-dimensional metric and α, together with an additional scalar field, are then recovered from the five-dimensional metric on $\tilde{M}$. The coupled Einstein-Maxwell equations describing gravitation and electromagnetism follow from the Einstein field equations on $\tilde{M}$.

This paper reviews the rich evolution of such models in physics with special regard to their interaction with mathematics. The original four plus one construction of $\tilde{M}$ later became recognized also as a bundle structure with the form α as a connection. In this case the Maxwell equations for α can be expressed in terms of the curvature form induced by α. The problem with one-dimensional fiber, S^1, and Abelian group U(1), naturally generalizes to higher dimensional spaces and groups, leading to what physicists refer to as non-Abelian Yang-Mills gauge theories. This paper reviews many interesting results relating the geometry of a space (e.g. that it be Einstein, with certain isometries) to the condition that it be the total space of a bundle with connection satisfying Yang-Mills equations and other conditions.　*C.H. Brans.*

Author profile:

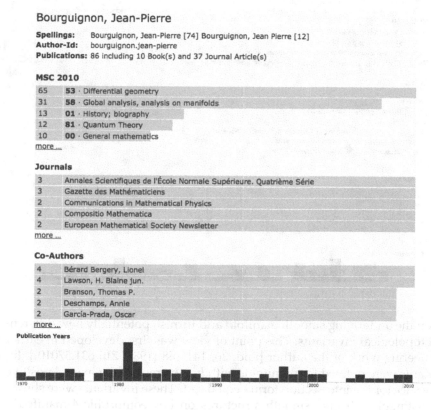

Bourguignon, Jean-Pierre

Spellings:　Bourguignon, Jean-Pierre [74] Bourguignon, Jean Pierre [12]
Author-Id:　bourguignon.jean-pierre
Publications: 86 including 10 Book(s) and 37 Journal Article(s)

MSC 2010

65	53 · Differential geometry
31	58 · Global analysis, analysis on manifolds
13	01 · History; biography
12	81 · Quantum Theory
10	00 · General mathematics

more ...

Journals

3	Annales Scientifiques de l'École Normale Supérieure. Quatrième Série
3	Gazette des Mathématiciens
2	Communications in Mathematical Physics
2	Compositio Mathematica
2	European Mathematical Society Newsletter

more ...

Co-Authors

4	Bérard Bergery, Lionel
4	Lawson, H. Blaine jun.
2	Branson, Thomas P.
2	Deschamps, Annie
2	García-Prada, Oscar

more ...

Publication Years

Zbl 715.57007
Donaldson, S.K.
Polynomial invariants for smooth four-manifolds.
Topology 29, No.3, 257-315 (1990).

The classical invariants for smooth manifolds which have been so successful for the classification of manifolds in higher dimensions have not been effective for the classification of smooth 4-manifolds. In the early 1980's the author introduced to the study of 4-manifolds dramatically new methods using Yang-Mills and gauge theory to obtain restrictions on the intersection forms of smooth 4-manifolds [J. Differ. Geom. 18, 279-315 (1983; Zbl 507.57010)]. These restrictions are not detected by classical invariants. The Yang-Mills equations depend on the Riemannian geometry of the underlying 4-manifold; however, certain homological properties of the moduli space of solutions are invariant under continuous change of metric. Since any two metrics can be joined by a path, these homological properties depend only

162

Author profile:

Donaldson, Simon K.

Spellings: Donaldson, S.K. [57] Donaldson, Simon K. [12] Donaldson, Simon [6] Donaldson, S. K. [4] Donaldson, S. [2]
Author-Id: donaldson.simon-k
Publications: 81 including 8 Book(s) and 42 Journal Article(s)

MSC 2010

52	**53** · Differential geometry
42	**57** · Manifolds and cell complexes
28	**58** · Global analysis, analysis on manifolds
18	**32** · Functions of several complex variables and analytic spaces
13	**81** · Quantum Theory

more ...

Journals

11	Journal of Differential Geometry
3	Topology
2	Bulletin (New Series) of the American Mathematical Society
2	Communications in Mathematical Physics
2	Geometric and Functional Analysis - GAFA

more ...

Co-Authors

3	Katzarkov, Ludmil
2	Auroux, Didier
2	Braam, Peter J.
2	Kronheimer, Peter Benedict
2	Thomas, Charles B.jun.

more ...

Publication Years

upon the underlying smooth manifold and furnish potentially new differential topological invariants. This point of view was first developed in earlier pioneering work of the author [ibid. 26, 141-168 (1987; Zbl 631.57010)]. Invariants were defined for manifolds with b_+^2 - the rank of a maximal positive subspace for the intersection form - equal to 1. These invariants were shown to distinguish distinct smooth structures on homeomorphic 4-manifolds. Later, R. Friedman and J. W. Morgan [ibid. 27, 297-398 (1988; Zbl 669.57016 and Zbl 669.57017)] and C. Okonek and A. Van de Ven [Invent. Math. 86, 357-370 (1986; Zbl 613.14018)] showed that these invariants can distinguish mutually distinct differentiable structures on an infinite family of homeomorphic 4-manifolds.

The paper under review extends these ideas. For simply-connected smooth 4-manifolds X with $b_+^2 > 1$ and odd, the author defines an infinite set of differential topological invariants (which have become known as Donaldson invariants) for X which are distinguished elements $q_{k,X}$, $k \in \mathbb{Z}$, of the ring $S^*(H_2(X))$ of polynomials in the homology of X. He then uses these invariants to obtain spectacular results concerning the differential topology of algebraic surfaces. The most striking results are:

Theorem A. Let S be a simply-connected, smooth, complex projective surface. If S is orientation preserving diffeomorphic to the connected sum of two oriented 4-manifolds, then one of these manifolds has a negative definite intersection form.

This follows immediately from two other theorems:

Theorem B. Suppose X is a simply-connected, oriented smooth 4-manifold with b_+^2 odd and X is orientation preserving diffeomorphic to the connected sum of two oriented 4-manifolds where the b_+^2 of both manifolds is strictly positive. Then all the Donaldson invariants for X vanish.

Theorem C. Let S be a simply-connected complex projective surface and H a hyperplane class in $H_2(S)$. Then for a large enough k, $q_{k,S}(H, \ldots, H) > 0$.

These results are impressive, as are their proofs. The proofs are an artful and demanding blend of differential and algebraic topology and geometry and local and global analysis. This paper is a mine for technique and ideas and, together with the author's paper in J. Differ. Geom. 24, 275-341 (1986; Zbl 635.57007) and the new book by P. Kronheimer and the author [The geometry of four-manifolds, Oxford: Clarendon Press (1990; Zbl 820.57002)], make for indispensible resources for the modern study of smooth 4-manifolds.

R. Stern.

Zbl 756.42014
Bourgain, J.
Besicovitch type maximal operators and applications to Fourier analysis.
Geom. Funct. Anal. 1, No.2, 147-187 (1991).

Averaging over tubes of unit length and width $\delta < 1$ leads to two maximal functions f_δ^*, f_δ^{**}. The first is defined on the sphere S^{d-1} in $\mathbb{R}^d$ and for a fixed direction one takes the maximum over all translates of the tube (the Kakeya maximal function). The second is defined on $\mathbb{R}^d$ and at a given point x one considers all δ tubes centered at x and varying direction (the Nikodým maximal function). It is a natural conjecture that the L^p bound on f_δ^*, f_δ^{**} is given essentially by the formula $(1/\delta)^{d/p-1+\varepsilon}$ for $1 \le p \le d$. This fact is known for $d = 2$ but open in higher dimensions. In particular, it seems unknown whether a set in $\mathbb{R}^3$ containing a line in every direction needs to have full Hausdorff dimension. The estimate is verified here if $p < p(d)$, where $(d + 1)/2 < p(d) < d/2 + 1$. As an application the author proves that there are no so-called Besicovitch (d,k) sets provided that $d \le 2^{k-1+k}$. The author also applies these results to the restriction problem for the Fourier transform and spherical multipliers. M. Milman (Boca Raton)

Author profile:

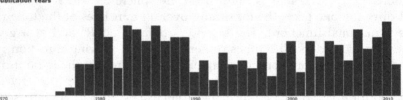
Zbl 776.42018
Daubechies, Ingrid
Ten lectures on wavelets.
CBMS-NSF Regional Conference Series in Applied Mathematics. 61. Philadelphia, PA: SIAM, Society for Industrial and Applied Mathematics. xix, 357 p. (1992).

Wavelet theory found many applications during the past decade. This book is one of the few recent books on wavelets. The authoress has made important contributions to the theory of wavelets. The book consists of an introduction, 10 chapters, a bibliography, a subject index and an author index.

In Chapter 1 the authoress explains why the wavelets are of interest. In Chapter 2 the notion and properties of the continuous wavelet transform are studied. In Chapter 3 the discrete wavelet transforms are studied. In Chapter 4 the role of time-frequency density in wavelet transforms is discussed. Necessary conditions for the set $\{\exp\{2\pi imx\}g(x-n)\} := \{g_{mn}(x)\}$ to form a frame are given. The Zak transform is briefly discussed. Orthonormal wavelet bases

Author profile:

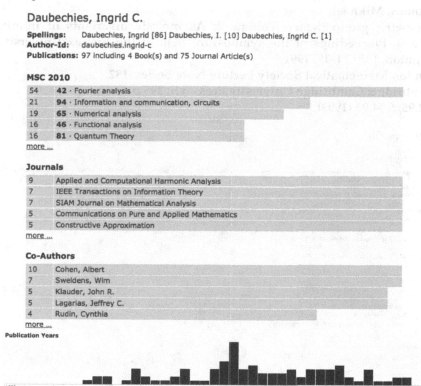

Daubechies, Ingrid C.

Spellings: Daubechies, Ingrid [86] Daubechies, I. [10] Daubechies, Ingrid C. [1]
Author-Id: daubechies.ingrid-c
Publications: 97 including 4 Book(s) and 75 Journal Article(s)

MSC 2010

54	**42** · Fourier analysis
21	**94** · Information and communication, circuits
19	**65** · Numerical analysis
16	**46** · Functional analysis
16	**81** · Quantum Theory

more ...

Journals

9	Applied and Computational Harmonic Analysis
7	IEEE Transactions on Information Theory
7	SIAM Journal on Mathematical Analysis
5	Communications on Pure and Applied Mathematics
5	Constructive Approximation

more ...

Co-Authors

10	Cohen, Albert
7	Sweldens, Wim
5	Klauder, John R.
5	Lagarias, Jeffrey C.
4	Rudin, Cynthia

more ...

Publication Years

are constructed. These bases have good localization properties in both time and frequency domains. In Chapter 5 multiresolution analysis is discussed and its connections with subband filtering and QMF filters are described. In Chapter 6 orthonormal bases of compactly supported wavelets are studied. Cohen's theorem is proved. Examples of compactly supported wavelets, generating an orthonormal basis, are given. In Chapter 7 the regularity of the compactly supported wavelets is discussed. In Chapter 8 symmetry and lack of it for compactly supported wavelet bases is discussed. Except the Haar basis, no symmetric or antisymmetric compactly supported real valued wavelet bases exist. Symmetric biorthogonal wavelet bases do exist. In Chapter 9 characterizations of the functional spaces $L^p(\mathbb{R})$, $H^s(\mathbb{R})$ (Sobolev spaces), $C^s(\mathbb{R})$ (Hölder spaces), $0 < s < 1$, are given in terms of the wavelet coefficients of a function belonging to one of these spaces. In Chapter 10 some generalizations and multidimensional wavelets are discussed, wavelets on an interval are studied, and the "splitting trick" is described.

A.G. Ramm (Manhattan)

Zbl 841.20039
Gromov, Mikhael
Geometric group theory. Volume 2: Asymptotic invariants of infinite
groups. Proceedings of the symposium held at the Sussex University,
Brighton, July 14–19, 1991
London Mathematical Society Lecture Note Series. 182.
Cambridge: Cambridge University Press. vii, 295 p.
£22.95; $ 34.95 (1993).

Author profile:

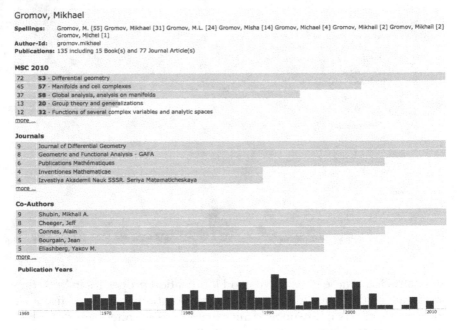

Gromov, Mikhael

Spellings: Gromov, M. [55] Gromov, Mikhael [31] Gromov, M.L. [24] Gromov, Misha [14] Gromov, Michael [4] Gromov, Mikhail [2] Gromov, Mikhaïl [2] Gromov, Michel [1]
Author-Id: gromov.mikhael
Publications: 135 including 15 Book(s) and 77 Journal Article(s)

MSC 2010

72	**53** · Differential geometry
45	**57** · Manifolds and cell complexes
37	**58** · Global analysis, analysis on manifolds
13	**20** · Group theory and generalizations
12	**32** · Functions of several complex variables and analytic spaces

more…

Journals

9	Journal of Differential Geometry
8	Geometric and Functional Analysis - GAFA
6	Publications Mathématiques
4	Inventiones Mathematicae
4	Izvestiya Akademii Nauk SSSR. Seriya Matematicheskaya

more…

Co-Authors

9	Shubin, Mikhail A.
8	Cheeger, Jeff
6	Connes, Alain
5	Bourgain, Jean
5	Eliashberg, Yakov M.

more…

Publication Years

This ground breaking paper (a book) may be enthusiastically recommended
to researchers in several subjects – combinatorial group theory, geometry
and topology of manifolds of non-positive curvature, dynamics, etc. It provides a fascinating look at finitely generated groups, which continues the
author's study of infinite groups from a geometric viewpoint started in his
previous works [Ann. Math. Stud. 97, 183-213 (1981; Zbl 467.53035); Proc.
ICM, Warsawa 1983, Vol. I, 385-392 (1984; Zbl 599.20041), MSRI Publ., Math.
Sci. Res. Inst. 8, 75-263 (1987; Zbl 634.20015), see also *D. Epstein, J. Cannon, D.
Holt, S. Levy, M. Paterson* and *W. Thurston*, Word processing in groups (1992;
Zbl 764.20017)]. The author has again opened up new vistas that geometers
and geometric group theorists may explore in the years to come.
The major section titles are: (1) Asymptotic methods (thickening, asymp-
totic connectedness, contractibility, large scale dimension, (co)homology and
width); (2) Asymptotic cones (in particular, for Lie groups and lattices; ge-

ometry of such cones and the space of the word metrics); (3) Distorsion (with intrinsic geometry of subgroups and submanifolds; polynomial, exponential and nonrecursive distortion); (4) Topology of balls (indiscrete and highly non-contractible in the conventional sense spaces); (5) Filling invariants (isoperimetric and isodiametric functions, area filling in nilpotent and solvable groups, filling length, radius and Morse landscape, filling on the large scale and volume distortion of subspaces); (6) Semi-hyperbolic spaces ($K \leq 0$, coning and combing, rank and geometry of flags, Tits boundary and Morse landscape at infinity, minimal surfaces and norms on homology, group actions with $K \leq 0$); (7) Hyperbolic groups (groups and curvature, hierarchy in the hyperbolic ranks, pinching and conformal geometry at infinity, round trees, Kähler and anti-Kähler groups, uniform embeddings and T-amenability); (8) L_p-Cohomology (with Appendix on conformal hyperbolic groups and spaces); (9) Finitely presented groups. This list of headings hardly indicates the breadth of discussion.   *B.N. Apanasov (Norman)*

Zbl 867.57029
Witten, Edward
Monopoles and four-manifolds
Math. Res. Lett. 1, No.6, 769-796 (1994).

Motivated by his joint work with N. Seiberg on $N = 2$ supersymmetric Yang-Mills theory in 4 dimensions, the author gives here a set of monopole equations dual to the antiself-duality equations which have since become known as the Seiberg-Witten equations. He discusses how they arise physically, formulates the mathematical properties of their solutions and defines new invariants using them, relates these invariants to Donaldson invariants, and proves a number of results concerning them. These new invariants have since revolutionized 4-manifold topology as they have led to simpler approaches to results previously proved using Donaldson invariants as well as providing a tool to prove many new results. One of the major breakthroughs in Donaldson theory due to *P. B. Kronheimer* and *T. S. Mrowka* [J. Differ. Geom. 41, No. 3, 573-734 (1995; Zbl 842.57022)] shortly before these new invariants were discovered was the reformulation of the Donaldson invariants in terms of a finite number of basic classes that occur in a formula for the Donaldson series for a 4-manifold of simple type. The author gives the sketch of a physical argument that these basic classes as well as their coefficients are given through the new Seiberg-Witten invariants. Although his conjecture has been verified in a large number of examples, a full proof still remains a major topic of current work. Independent of this conjecture, these new invariants have provided an extremely powerful tool for studying 4-manifolds that has proved easier to use than the Donaldson invariant.
We describe here briefly the Seiberg-Witten equations and corresponding invariants. Let X be an oriented, closed 4-manifold with a Riemannian metric g. One may always choose a Spin^c structure ξ on (X, g), which may be described as a principal Spin$^c(4)$ bundle $\widetilde{P}$ covering the orthogonal frame bundle P over

168

Author profile:

Witten, Edward

Spellings: Witten, Edward [165] Witten, E. [19]
Author-Id: witten.edward
Publications: 184 including 4 Book(s) and 129 Journal Article(s)

MSC 2010

171	81 · Quantum Theory
44	53 · Differential geometry
43	83 · Relativity and gravitational theory
37	58 · Global analysis, analysis on manifolds
31	14 · Algebraic geometry

more ...

Journals

32	Nuclear Physics. B
17	Journal of High Energy Physics [electronic only]
16	Communications in Mathematical Physics
13	Advances in Theoretical and Mathematical Physics
6	International Journal of Modern Physics A

more ...

Co-Authors

9	Seiberg, Nathan
7	Gukov, Sergei
7	Vafa, Cumrun
4	Klebanov, Igor R.
3	Friedman, Robert M.

more ...

Publication Years

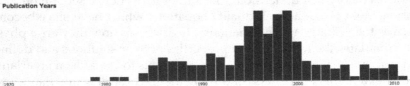

| 1970 | 1980 | 1990 | 2000 | 2010 |

X. Here $\mathrm{Spin}^c(4) = \mathrm{Spin}(4) \times S^1/\{\pm 1\}$ covers $\mathrm{SO}(4) = \mathrm{Spin}(4)/\{\pm 1\}$ on the fiber. Associated to this Spin^c structure are $U(2)$ bundles of spinors $S^\pm$ and their common determinant line bundle L. The data for the Seiberg-Witten equations is a pair (A, ψ), where A is a connection on L and ψ is a section of S^+. There is an identification of the traceless endomorphisms of S^+ with the self-dual 2-forms, through which the section ψ determines a 2-form $q(\psi)$ using the trace-free part of the endomorphism $\psi \otimes \psi^*$. The Seiberg-Witten equations are

$$F_A^+ = q(\psi), \quad D_A \psi = 0.$$

Here D_A is the Dirac operator formed using the connection on $\widetilde{P}$ covering the Levi-Civita connection on P and the connection A on L. There is a gauge group given by the automorphisms of $\widetilde{P}$ which cover the identity automorphism of P, which can be identified with maps from X to S^1, and one can form the moduli space $\mathcal{M}$ of the gauge equivalence classes of solutions to these equations. As usual, the proper analytic formulation requires the use of appropriate Sobolev spaces. The virtual dimension of this moduli space is given by the index theorem to be $d = \frac{1}{4}(c_1(L)^2 - (2\chi + 3\sigma))$. To get a manifold of this dimension requires adding a small perturbation term ih to the right hand side of the first equation as well as a condition $b_2^+ \geq 1$ needed to generically avoid reducible solutions (those where $\psi = 0$). In particular, there will generically be no solutions when $d < 0$.

In contrast to Donaldson theory where the lack of compactness causes many technical difficulties, here $\mathcal{M}$ is shown to be compact through use of the Weitzenböck formula for the Dirac operator. This formula plays a special role in a number of other results, including the fact that for a fixed metric there are only a finite number of Spin^c structures where the moduli space is nonempty, and for a positive scalar curvature metric there are only reducible solutions (and none of those generically if $b_2^+ \geq 1$). As in Donaldson theory, one uses a homology orientation of $H^1(X; \mathbb{R}) \oplus H_+^2(X; \mathbb{R})$ to orient the moduli space when it is a manifold. When $d = 0$ (which occurs for example if the Spin^c structure comes from an almost complex structure on the tangent bundle) the oriented moduli space will consist of a finite number of signed points. In this case the Seiberg-Witten invariant SW(ξ) is defined to be the sum of the signs at these points. When $d > 0$ and is even, the first Chern class μ of the restriction of the base point fibration to $\mathcal{M}$ is used to define SW$(\xi) = \int_{\mathcal{M}} \mu^{d/2}$. The invariant is defined to be 0 when d is odd. As long as $b_2^+ > 1$, reducible connections can be avoided in paths (g_t, h_t) of metric-perturbation pairs, and a cobordism argument gives that SW(ξ) is a diffeomorphism invariant independent of generic (g, h). In particular, all Seiberg-Witten invariants must vanish if X has a metric of positive scalar curcature. When $b_2^+ = 1$, the paper gives a wall crossing formula which shows how the invariant changes when a path goes through a wall corresponding to a reducible connection when $b_1 = 0$. More general wall crossing formulas have since been given by *T. J. Li* and *A. Liu* [General wall crossing formula, Math. Res. Lett. 2, No. 6, 797-810 (1995; Zbl 871.57017)] and play an important role in the analysis of the case $b_2^+ = 1$.

A class L with non-vanishing Seiberg-Witten invariant is called a basic class – there is a physically based argument that the basic classes with $d = 0$ are the same as in Donaldson theory. For $b_2^+ > 1$, the basic classes occur in pairs as the conjugate Spin^c structure $\bar{\xi}$ has SW$(\bar{\xi}) = \pm$SW(ξ). Besides establishing the main properties of the invariant, a method is given for computing it for Kähler manifolds where the invariant is given a holomorphic reformulation. There it is shown that the invariant can only be nonzero when $d = 0$, and for the Spin^c structures determined by $\pm K$ with K the canonical class one has SW$(\pm K) = \pm 1$.

It is also shown that for minimal surfaces of general type the only basic classes are $\pm K$, proving that the canonical class is a diffeomorphism invariant of these surfaces. Another major result in the paper is that the Seiberg-Witten invariant vanishes whenever $X = X_1 \# X_2$ with $b_2^+(X_i) \geq 1$. There has been a lot of important work that has been based on the Seiberg-Witten invariants introduced in this paper, and we want to inform the reader of a few of the more important papers that have appeared. *P. B. Kronheimer* and *T. S. Mrowka*

[Math. Res. Lett. 1, No. 6, 797-808 (1994; Zbl 851.57023)] gave a proof of the Thom conjecture which states that a holomorphic curve in $\mathbb{C}P^2$ minimizes the genus of any surface which represents the same homology class. They had earlier proved forms of the generalized Thom conjecture and given a generalized adjunction inequality for other manifolds using Donaldson theory, but these proofs did not work for $\mathbb{C}P^2$. Their earlier results can be reproven using Seiberg-Witten theory. *J. W. Morgan, M. G. Szabo* and *C. H. Taubes* [J. Differ. Geom. (to appear)] have proven that a smooth symplectic curve of nonnegative self-intersection in a symplectic four-manifold is genus minimizing, and have given a product formula for computing Seiberg-Witten invariants which allows them to give a non-vanishing result for the Seiberg-Witten invariants of certain generalized connected sum manifolds. *C. H. Taubes* has published a number of papers [Math. Res. Lett. 1, No. 6, 809-822 (1994; Zbl 853.57019); ibid. 2, No. 1, 9-13 (1995; Zbl 854.57019); ibid. 2, No. 2, 221-238 (1995; Zbl 854.57020); SW $\Rightarrow$ Gr: From the Seiberg-Witten equations to pseudo-holomorphic curves, J. Am. Math. Soc. 9, No. 3, 845-918 (1996; Zbl 867.53025)] which have generalized results for Kähler surfaces to symplectic four-manifolds, and identified Seiberg-Witten invariants with Gromov invariants. *C. LeBrun* has found a number of differential geometric applications of the Seiberg-Witten invariants, including proving uniqueness results for Einstein metrics for smooth compact quotients of complex hyperbolic 2-space ["Einstein metrics and Mostow rigidity," Math. Res. Lett. 2, No. 1, 1–8 (1995; Zbl 974.53035)] and finding infinitely many compact simply connected smooth four-manifolds which do not admit Einstein metrics but still satisfy the strict Hitchin-Thorpe inequality [ibid. 3, No. 2, 133-147 (1996; Zbl 856.53035)]. Furuta has used the Seiberg-Witten moduli space to prove a slightly weaker version of the 11/8 conjecture. There are also now available expository treatments of the Seiberg-Witten invariants, including a paper by *S. K. Donaldson* ["The Seiberg-Witten equations and 4-manifold topology," Bull. Am. Math. Soc. 33, No. 1, 45–70 (1996; Zbl 872.57023)] and a book by *J. W. Morgan* [The Seiberg-Witten equations and applications to the topology of smooth four-manifolds, Princeton Math. Notes 44 (1996; Zbl 846.57001)].   *T. Lawson (New Orleans)*

Zbl 823.11029
Wiles, Andrew
Modular elliptic curves and Fermat's Last Theorem.
Ann. Math. (2) 141, No. 3, 443-551 (1995).

The main result is the proof of the Taniyama-Weil conjecture for a large class of elliptic curves over $\mathbb{Q}$. These include semistable curves, and thus the result implies the famous Fermat conjecture.

To achieve this one shows that in many cases the Hecke algebra of a modular curve is the base of a universal deformation of the associated p-adic Galois representation. Here $p \geq 3$, and the representation modulo p must be irreducible. If this holds for $p = 3$, then everything follows from results of

Author profile:

Wiles, Andrew J.

Spellings: Wiles, A. [13] Wiles, Andrew [8] Wiles, A.J. [3] Wiles, Andrew J. [1]
Author-Id: wiles.andrew-j
Publications: 25 including 1 Book(s) and 19 Journal Article(s)

MSC 2010

24	11 · Number theory
12	14 · Algebraic geometry
2	13 · Commutative algebra
1	00 · General mathematics
1	01 · History; biography

Journals

6	Annals of Mathematics. Second Series
4	Inventiones Mathematicae
2	Duke Mathematical Journal
1	American Journal of Mathematics
1	Annales de la Faculté des Sciences de Toulouse. Série VI. Mathématiques

more ...

Co-Authors

4	Coates, John H.
4	Skinner, Christopher M.
3	Mazur, Barry
1	Carlson, Jordan
1	Çiperiani, Mirela

more ...

Publication Years

Langlands-Tunnell, as $PGL(2, \mathbb{F}_3) \cong S_4$ is solvable. If the mod 3 representation is reducible, one can use $p = 5$ (and the result for $p = 3$).

In the meantime there has been more progress, extending the result to elliptic curves with semistable reduction at 3 and 5. The restriction stems from the argument above, and limitations of our present crystalline techniques.

The contents in more detail.

Chapter I introduces the universal deformation ring, various local conditions on representations, and the corresponding tangent spaces. These are H^1's of certain cohomology theories, and the corresponding Euler characteristic is computed using the results of Tate-Poitou.

Chapter II treats Hecke algebras. It is shown that they are Gorenstein (this comes down to multiplicity one), and Ribet's theory of change of level is used to start the reduction to the minimal case. A key fact is always that certain Hecke operators are redundant in the definition of the Hecke algebra. This is easy for primes of good reduction, but involved for the others.

Chapter III brings the introduction of certain auxiliary primes $q \equiv 1 \bmod p^n$ which are also very important in the subsequent paper of Taylor-Wiles [Ann. Math. (2) 141, No. 3, 553–572 (1995; Zbl 823.11030)]. It then reduces the assertion to the fact that the Hecke algebra is a complete intersection. That this condition holds is the content of Taylor-Wiles.

Chapter IV treats the dihedral case. This does not occur for semistable curves, and requires the techniques of Kolyvagin-Rubin.

Chapter V actually proves the Taniyama-Weil conjecture for many elliptic curves.

An appendix explains the relevant commutative algebra. *G. Faltings (Bonn)*

Zbl 823.11030
Taylor, Richard; Wiles, Andrew
Ring-theoretic properties of certain Hecke algebras.
Ann. Math. (2) 141, No. 3, 553-572 (1995).

Author profile:

Taylor, Richard

Spellings: Taylor, Richard [41] Taylor, R. [11]
Author-Id: taylor.richard
Publications: 52 including 1 Book(s) and 41 Journal Article(s)

MSC 2010

32	11 · Number theory
11	14 · Algebraic geometry
8	05 · Combinatorics
2	22 · Topological groups, Lie groups
2	68 · Computer science

more ...

Journals

5	Inventiones Mathematicae
4	Journal of the American Mathematical Society
3	Duke Mathematical Journal
3	Proceedings of the Physical Society
2	Annals of Mathematics. Second Series

more ...

Co-Authors

5	Diamond, Fred
4	Harris, Michael
2	Buzzard, Kevin
2	Caldwell, Brian D.
2	Conrad, Brian

more ...

Publication Years

This paper provides a key fact needed in the previous paper by *A. Wiles* [Ann. Math. (2) 141, No. 3, 443–551 (1995; Zbl 823.11029); namely that (certain) Hecke algebras are complete intersections. The key ideas are a computation of Euler characteristics (Tate-Poitou) and the introduction of auxiliary primes $p \equiv 1 \bmod p^n$ such that the projective limit (as $n \to \infty$) of the corresponding Hecke algebras is a power series ring. To see this one estimates its minimal number of generators (say d) and then shows that its dimension is at least $d + 1$. To derive the assertion one uses that the original Hecke algebra can be defined by dividing this projective limit by d equations. It should be noted that this treats only the minimal case, where the condition "minimal" is used in the computation of Euler characteristics.

An appendix explains a remark of the reviewer which allows one to simplify some arguments. The authors were probably too exhausted to find this additional shortcut.　　　　　　　　　　　　　　　　　　　　　　　　 *G. Faltings (Bonn)*

Zbl 873.14039
Rapoport, M.; Zink, Th.
Period spaces for p-divisible groups.
Annals of Mathematics Studies. 141. Princeton, NJ: Princeton Univ. Press.
xxi, 324 p. \$ 24.95; £24.00/pbk, \$ 59.50; £50.00/hbk (1996).

Let E be a p-adic field, and let Ω_E^d be the completion of all E-rational hyperplanes in the projective space $\mathbb{P}^d$. This is a rigid-analytic space over E equipped with an action of $GL_d(E)$. V. G. *Drinfel'd* [Funct. Anal. Appl. 10, 107-115 (1976); translation from Funkts. Anal. Prilozh. 10, No. 2, 29-40 (1976; Zbl 346.14010)] has constructed a system of unramified coverings $\widetilde{\Omega}_E^d$ of Ω_E^d to which the action of $GL_d(E)$ is lifted. Drinfeld has shown that these covering spaces can be used to p-adically uniformize the rigid-analytic spaces corresponding to Shimura varieties associated to certain unitary groups. Also Drinfeld has conjectured that the ℓ-adic cohomology group with compact supports $H_e^i(\widetilde{\Omega}_E^d \otimes E, \overline{\mathbb{Q}}_\ell)$ ($\ell \neq p$) should give a realization of all supercuspidal representations of $GL_d(E)$.

Author profile:

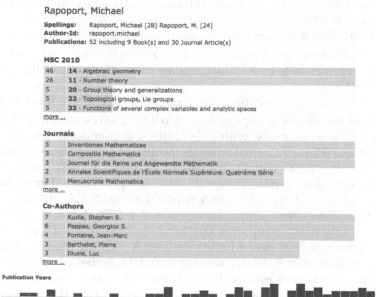

Rapoport, Michael

Spellings: Rapoport, Michael [28] Rapoport, M. [24]
Author-Id: rapoport.michael
Publications: 52 including 9 Book(s) and 30 Journal Article(s)

MSC 2010

46	14 · Algebraic geometry
26	11 · Number theory
5	20 · Group theory and generalizations
5	22 · Topological groups, Lie groups
5	32 · Functions of several complex variables and analytic spaces

more ...

Journals

5	Inventiones Mathematicae
3	Compositio Mathematica
3	Journal für die Reine und Angewandte Mathematik
2	Annales Scientifiques de l'École Normale Supérieure. Quatrième Série
2	Manuscripta Mathematica

more ...

Co-Authors

7	Kudla, Stephen S.
6	Pappas, Georgios S.
4	Fontaine, Jean-Marc
3	Berthelot, Pierre
3	Illusie, Luc

more ...

Publication Years

1970	1980	1990	2000	2010

In this monograph, the authors generalize Drinfeld's construction and results to other p-adic groups. Their construction is based on the moduli theory of p-divisible groups of a fixed isogeny type. The moduli spaces constructed here are formal schemes over the ring of integers O_E whose generic fibers yield rigid-analytic spaces generalizing Drinfeld's Ω_E^d, and the covering spaces are obtained by trivializing the Tate modules of the universal p-divisible groups over these formal schemes. Moreover, the authors show that these spaces may be used to uniformize the rigid-analytic spaces associated to general Shimura varieties. Also the authors exhibit a rigid-analytic period map from the covering spaces to one of the p-adic symmetric spaces associated to the p-adic group. – The main results are presented in the following manner: First the moduli problems of p-divisible groups and the representability theorem (which yields the formal schemes generalizing $\widetilde{\Omega_E^d}$) are described. Then the covering spaces and the rigid-analytic period morphisms are described. Finally the non-archimedean uniformization theorem for Shimura varieties is proved. The main results are now described in more detail.

The moduli problem of p-divisible groups is divided into two types:

(EL): this type parametrizes p-divisible groups with endomorphisms and level structures within a fixed isogeny class;

(PEL): this type parametrizes p-divisible groups with polarizations, endomorphisms and level structures within a fixed isogeny class.

Let p be a prime ($\neq 2$ for most results). Let L be an algebraically closed field of characteristic p, $W(L)$ be the ring of Witt vectors over L, and $K_0 = K_0(L) = W(L) \otimes_{\mathbb{Z}} \mathbb{Q}$, and let σ be the Frobenius automorphism of K_0. Denote by $\mathrm{Nilp}_{W(L)}$ the category of locally Noetherian schemes over $\mathrm{Spec}W(L)$ such that the ideal sheaf $p \cdot O_{W(L)}$ is locally nilpotent. Let $\mathfrak{X}$ be a p-divisible group over $\mathrm{Spec}L$. Let $\mathcal{M}$ be a functor, which associates to $S \in \mathrm{Nilp}W(L)$ the set of isomorphism classes of pairs (X, ρ) consisting of a p-divisible group X over S and a quasi-isogeny $\rho : \mathfrak{X} \times_{\mathrm{Spec}L} \overline{S} \to X \times_S \overline{S}$ of p-divisible groups over $\overline{S}$. Then $\mathcal{M}$ is representable by a formal scheme locally formally of finite type over $\mathrm{Spf}W(L)$.

This result allows one to establish the representability of the functors (denoted $\breve{\mathcal{M}}$) of p-divisible groups endowed with endomorphisms and level structures (EL), respectively, with polarizations and endomorphisms and level structures (PEL), by formal schemes locally formally of finite type over $\mathrm{Spf}O_{\breve{E}}$. (Here $\breve{E} = E \cdot K_0$ with $O_{\breve{E}}$ the ring of integers.) – These functors are shown to depend on certain "rational" and "integral" data. For instance, the "rational" data of type (EL) consists of a 4-tuple (B, V, b, μ), where B is a finite dimensional semi-simple $\mathbb{Q}_p$-algebra, V a finite left B-module. Let $G = GL_B(V)$ (algebraic group over $\mathbb{Q}_p$). Let $b \in G(K_0)$, and $\mu : \mathbb{G}_m \to G_K$ a homomorphism defined over a finite extension K/K_0. One requires that the filtered isocrystal over K, $(V \otimes_{\mathbb{Q}_p} K_0, b(\mathrm{id} \otimes \sigma), V_K^\bullet)$, is that of associated to a p-divisible group over $\mathrm{Spec}O_K$. For each pair (G, b) as above, there is the group $J(\mathbb{Q}_p)$ of quasi-isogenies of $\mathfrak{X}$. "Integral" data of type (EL) consists of

Author profile:

Zink, Thomas

Spellings: Zink, Thomas [25] Zink, Th. [5] Zink, T. [2]
Author-Id: zink.thomas
Publications: 32 including 2 Book(s) and 22 Journal Article(s)

MSC 2010

29	14 · Algebraic geometry
7	11 · Number theory
4	32 · Functions of several complex variables and analytic spaces
2	13 · Commutative algebra
2	94 · Information and communication, circuits

more ...

Journals

6	Mathematische Nachrichten
3	Beiträge zur Algebra und Geometrie
3	Documenta Mathematica
1	Annales Scientifiques de l'École Normale Supérieure. Quatrième Série
1	Annals of Mathematics. Second Series

more ...

Co-Authors

4	Langer, Andreas
3	Fitzner, Heinz-Jörg
3	Kleinert, Werner
3	Kurke, Herbert
3	Pfister, Gerhard

more ...

Publication Years

a maximal order O_B and an O_B-lattice chain $\mathcal{L}$ in V. Similarly "rational" and "integral" data of type (PEL) are defined. – Several examples of the formal schemes $\breve{\mathcal{M}}$ are given.

Next, the period morphism associated to the moduli problem $\breve{\mathcal{M}}$ of (EL) or (PEL)-type is described. Let B, V and G be as above. Let $\breve{\mathcal{M}}^{\mathrm{rig}}$ be the rigid-analytic space over $\breve{E}$ associated to $\breve{\mathcal{M}}$ (the generic fibre of $\breve{\mathcal{M}}$). Then the period morphism is defined as a rigid-analytic morphism from $\breve{\mathcal{M}}$ to $\breve{\mathcal{F}}^{\mathrm{rig}} = \mathcal{F} \times_{\mathrm{Spec}\,E} \breve{E}$, where $\mathcal{F}$ is the homogeneous projective algebraic variety under $G_{\breve{E}}$ defined by the conjugacy class of the one-parameter subgroup μ. – Further, properties of the period morphisms are discussed, e.g., $\breve{\pi}$ is étale and $J(\mathbb{Q}_p)$-equivariant, and the descriptions of the image of $\breve{\pi}$ and the associated Tate modules.

Finally, a non-archimedean uniformisation theorem for certain Shimura varieties is proved. Let B stand for a finite dimensional algebra over $\mathbb{Q}$ equipped with a positive anti-involution $*$, V a finite left B-module with a non-degenerate alternating bilinear form $(,)$ with values in $\mathbb{Q}$ satisfying certain conditions and $G = GL_B(V)$ (algebraic group over $\mathbb{Q}$). Then one obtains a Shimura variety over $E \subset \mathbb{C}$ associated to $(G, h : \mathrm{Res}_{\mathbb{C}/\mathbb{R}}\mathbb{G}_m\mathbb{G}_\mathbb{R})$. Fix data of type (PEL), i.e. $(B \otimes \mathbb{Q}_p, {}^*, V \otimes \mathbb{Q}_p, (,), b, \mu, O_B \otimes \mathbb{Z}_p, \Lambda)$. Here an order O_B of B is chosen so that $O_B \otimes \mathbb{Z}_p$ is a maximal order of $B \otimes_\mathbb{Q} \mathbb{Q}_p$ stable under $*$, and a self-dual $O_B \otimes_\mathbb{Z} \mathbb{Z}_p$-lattice Λ in $V \otimes_\mathbb{Q} \mathbb{Q}_p$. Fix an open compact subgroup $C^p \subset G(\mathbb{A}_f^p)$. These data define a moduli problem of (PEL) type parametrizing triples $(A, \overline{\lambda}, \overline{\eta}^p)$ consisting of an O_B-abelian variety A, a $\mathbb{Q}$-homogeneous principal O_B-polarization $\overline{\lambda}$, and a C^p-level structure $\overline{\eta}^p$ and which is representable by a quasi-projective scheme $\mathcal{A}_{C^p}$ over $\mathrm{Spec} O_{E_v}$. The generic fiber of $\mathcal{A}_{C^p}$ contains the Shimura variety (G, h) as an open and closed subscheme. Now fix a point $(A_0, \overline{\lambda}_0, \overline{\eta}_0^p)$ of $\mathcal{A}_{C^p}(L)$ and assume that it is basic. Then the set of points $(A, \overline{\lambda}, \overline{\eta}^p)$ of $\mathcal{A}_{C^p}(L)$ such that $(A, \overline{\lambda})$ is isogeneous to $(A_0, \overline{\lambda}_0)$ is a closed subset Z of $\mathcal{A}_{C^p}$. If $\mathcal{A}_{C^p}|Z$ denotes the formal completion of $\mathcal{A}_{C^p}$ along Z, then there is an isomorphism of formal schemes over $\mathrm{Spf} O_{E_v}$: $I(\mathbb{Q}) \backslash [\mathcal{M} \times G(\mathbb{A}_f^p)/C^p] \sim \mathcal{A}_{C^p}|Z$. (Here $I(\mathbb{Q})$ is the group of quasi-isogenies of $(A_0, \overline{\lambda}_0)$ that acts diagonally through suitable embeddings $I(\mathbb{Q}) \to J(\mathbb{Q}_p); I(\mathbb{Q}) \to G(\mathbb{A}_f^p)$.)

The contents of this monograph is as follows: §1: p-adic symmetric domains, §2: Quasi-isogenies of p-divisible groups, §3: Moduli spaces of p-divisible groups (with Appendix: Normal forms of lattice chains), §4: The formal Hecke correspondence, §5: The period morphism and the rigid-analytic coverings, §6: The p-adic uniformization of Shimura varieties, and bibliography and index.   N. Yui (Kingston/Ontario)

Zbl 899.11026
Lafforgue, Laurent
Drinfeld's 'shtukas' and the Ramanujan-Petersson conjecture.
Astérisque. 243. Paris: Société Mathématique de France, 329 p. (1997).

The author extends a number of results of Drinfeld concerning the Langlands correspondence for the GL(2)-case to the higher rank situation. The main theorem he obtains in doing this is a large part of the Ramanujan-Petersson conjecture for the function field case. The main tool for proving such results are generalizations of Drinfeld's concept of the so-called "shtukas" to the case of higher rank vector bundles by combining it with the concept of $\mathcal{D}'$-elliptic sheaves of the paper of G. Laumon, M. Rapoport and U. Stuhler [Invent. Math. 113, 217-338 (1993; Zbl 0809.11032)] by considering an additional action of a sheaf of division algebras, which makes "things more compact", at least for the case $r = 1$.

The first three chapters of the book contain a study of $\mathcal{D}'$-shtukas of arbitrary rank. Chapter I proves representability of the functor of rank r $\mathcal{D}'$-shtukas with level structure as an algebraic stack $\mathrm{Cht}^r_{\mathcal{D}', I}$. Chapter II studies questions of reducibility of $\mathcal{D}'$-shtukas which occurs only for rank $r \geq 2$ and lead to some interesting modifications of the Harder-Narasimhan filtration of vector bundles on curves. Chapter III studies these concepts over finite fields and gives the adelic description of shtukas which ultimately comes down

Author profile:

to the description of a vector bundle as a system of lattices. Chapter IV studies the special case of rank $r = 1$, which is easier in several respects. Finally, in chapters V and VI, the calculation of Lefschetz numbers (traces of operators of type (Frobenius)×(Hecke) on cohomology) and the evaluation of the Arthur-Selberg trace formula and its application to the Ramanujan-Petersson conjecture are done.   *U. Stuhler (Göttingen)*

Zbl 914.57015
Fintushel, Ronald; Stern, Ronald J.
Knots, links, and 4-manifolds.
Invent. Math. 134, No.2, 363-400 (1998).

This paper gives a construction which allows the authors to construct from a smooth 4-manifold X satisfying certain hypotheses and a knot K in the 3-sphere a new 4-manifold X_K which is homeomorphic to X but which can be distinguished smoothly from X by the Seiberg-Witten invariant whenever the Alexander polynomial of the knot is non-trivial. X_K is constructed from X by gluing in the knot complement crossed with a circle to the complement of a trivial normal bundle of torus T in X.

The required technical conditions on X are that it is simply connected with $b_2^+ > 1$, that T lives in a cusp neighborhood in X (such as the neighborhood of a cusp fiber in an elliptic surface), and that the complement of T in X is simply connected. The formal relationship of the Seiberg-Witten invariants of X and X_K is $\mathcal{SW}_{X_K} = \mathcal{SW}_X \cdot \Delta_K(t)$. Here $\mathcal{SW}_K$ is regarded as a finite

178

Author profile:

Fintushel, Ronald

Spellings: Fintushel, Ronald [55] Fintushel, R. [1]
Author-Id: fintushel.ronald
Publications: 56 including 47 Journal Article(s)

MSC 2010

53	57 · Manifolds and cell complexes
11	53 · Differential geometry
8	55 · Algebraic topology
7	14 · Algebraic geometry
4	81 · Quantum Theory

more ...

Journals

6	Journal of Differential Geometry
4	Pacific Journal of Mathematics
3	Algebraic & Geometric Topology
3	Annals of Mathematics. Second Series
3	Inventiones Mathematicae

more ...

Co-Authors

41	Stern, Ronald Jay
3	Pao, Peter Sie
1	Edmonds, Allan L.
1	Edmonds, Allen L.
1	Lawson, Terry C.

more ...

Publication Years

Laurent polynomial with variables being exponentials of the basic classes (its symmetric form using the fact that $-\beta$ is a basic class whenever β is with corresponding value), and $\Delta_K(t)$ is the symmetric Alexander polynomial with $\Delta_K(t^{-1}) = \Delta_K(t)$. Moreover, when the knot is fibered and X has a symplectic structure, the construction can be performed symplectically so that X_K has a symplectic structure. The authors use *C. H. Taubes*'s results [Math. Res. Lett. 1, No. 6, , 809-822 (1994; Zbl 853.57019); 2, No. 1, 9-13 (1995; Zbl 854.57019; No. 2, 221-238 (1995; Zbl 854.57020); J. Am. Math. Soc. 9, No. 3, 845-918 (1996; Zbl 867.53025)] concerning Seiberg-Witten and Gromov invariants of symplectic manifolds to give a similar relationship between Gromov invariants in the symplectic case as well as to show that if $\Delta_K(t)$ is not monic, then X_K does not possess a symplectic structure. The adjunction inequality is used to show that the existence of certain surfaces in X (which are readily found in many cases) will imply that X_K with the opposite orientation also has no symplectic structure. One case where these constructions apply is when X is the $K3$ surface with Seiberg-Witten invariant 1.

A corollary is that any A-polynomial $P(t)$ can occur as the Seiberg-Witten invariant of an irreducible homotopy $K3$ surface; if $P(t)$ is not monic, then the homotopy $K3$ surface does not admit a symplectic structure with either orientation; any monic A-polynomial can occur as the Seiberg-Witten invariant of a symplectic homotopy $K3$ surface. The authors also show that their new

179

Author profile:

Stern, Ronald Jay

Spellings: Stern, Ronald J. [100] Stern, R.J. [27] Stern, Ronald [4] Stern, R. J. [4] Stern, R. [2] Stern, Ronald Jay [1]
Author-Id: stern.ronald-jay
Publications: 138 including 14 Book(s) and 105 Journal Article(s)

MSC 2010

73	**57** · Manifolds and cell complexes
27	**49** · Calculus of variations and optimal control; optimization
25	**93** · Systems theory; control
24	**53** · Differential geometry
12	**00** · General mathematics

more ...

Journals

9	Linear Algebra and its Applications
7	Journal of Mathematical Analysis and Applications
6	Journal of Differential Geometry
5	Nonlinear Analysis. Theory, Methods & Applications. Series A: Theory and Methods
4	Annals of Mathematics. Second Series

more ...

Co-Authors

41	Fintushel, Ronald
16	Clarke, Franke H.
10	Akbulut, Selman
10	Önder, Turgut
9	Ledyaev, Yu.S.

more ...

Publication Years

examples cannot occur through a series of log transforms because of differing Seiberg-Witten invariants. The authors generalize their construction to the situation of n-component links where they form a fibered sum along the link components $\times S^1$ with n 4-manifolds X_i. They derive a formula for the Seiberg-Witten of the result and show it is the product of the Seiberg-Witten invariants of the fiber sums $E(1)\#_{F=T_j}X_j$ and the multivariable Alexander polynomial $\Delta_L(t_1,\ldots,t_n)$. In particular, the Seiberg-Witten invariant when all of the X_j are the rational elliptic surface $E(1)$ will just be $\Delta(t_1,\ldots,t_n)$. When L is a two component link with odd linking number, then the manifold $E(1)_L$ constructed from two copies of $E(1)$ will have polynomials which are not products of A-polynomials and will occur as Seiberg-Witten invariants of homotopy $K3$ surfaces. In particular, Z. *Szabó*'s examples $X(k)$, ($k \in \mathbb{Z}$, $k \neq 0,1$) [Invent. Math. 132, No. 3, 457-466 (1998; Zbl 906.57014)] of nonsymplectic simply connected irreducible smooth 4-manifolds occur as $E(1)_{W(k)}$, where $W(k)$ is the 2-component k-twisted Whitehead link, and the authors' examples $Y(k)$ of nonsymplectic $K3$ surfaces occur as $K3_{T(k)}$, where $T(k)$ is the k-twist knot. One critical tool for the computations in the paper are gluing theorems of *J. W. Morgan, T. S. Mrowka* and *Z. Szabó* [Math. Res. Lett. 4, No. 6, 915-929 (1997; Zbl 892.57021)], together with independent and joint work of these authors with Taubes that compute relative Seiberg-Witten invariants for c-embedded tori in terms of the absolute Seiberg-Witten invariant, as well as to compute Seiberg-Witten invariants of fiber sums and internal fiber

180

sums. The authors also use gluing formulas for generalized log transforms. These gauge theory tools are then combined with geometric results of J. Hoste [Pac. J. Math. 112, 347-382 (1984; Zbl 539.57004)] extending earlier work of W. R. Brakes [Lond. Math. Soc. Lect. Note Ser. 48, 27-37 (1982; Zbl 483.57009)] on sewn-up link exteriors. These results are combined to show that the effect of certain constructions given here on the Seiberg-Witten invariants satisfies the same axiomatic rules as occur in a computation of the Alexander polynomial due to Conway via skein relations, and this allows the authors to prove their main formulas on Seiberg-Witten invariants. *T. Lawson (New Orleans)*

Zbl 909.03048
Simpson, Stephen G.
Subsystems of second order arithmetic
Perspectives in Mathematical Logic.
Berlin: Springer. xiv, 444 p.
DM 98.00; öS 716.00; sFr. 89.50; £37.50; $ 59.95 (1999).

Author profile:

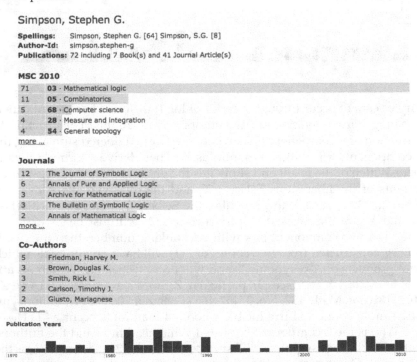

Simpson, Stephen G.

Spellings: Simpson, Stephen G. [64] Simpson, S.G. [8]
Author-Id: simpson.stephen-g
Publications: 72 including 7 Book(s) and 41 Journal Article(s)

MSC 2010

71	03 · Mathematical logic
11	05 · Combinatorics
5	68 · Computer science
4	28 · Measure and integration
4	54 · General topology

more ...

Journals

12	The Journal of Symbolic Logic
6	Annals of Pure and Applied Logic
3	Archive for Mathematical Logic
3	The Bulletin of Symbolic Logic
2	Annals of Mathematical Logic

more ...

Co-Authors

5	Friedman, Harvey M.
3	Brown, Douglas K.
3	Smith, Rick L.
2	Carlson, Timothy J.
2	Giusto, Mariagnese

more ...

Publication Years

The main question studied in the book under review is "What are the appropriate axioms for mathematics?" Since the author focuses on the language of second-order arithmetic, he is especially interested in the question of which set existence axioms are needed to prove the known theorems of mathematics. The attention is restricted to ordinary non-set-theoretic mathematics, i.e. to those branches of mathematics which are prior to or independent of the introduction of abstract set-theoretic concepts (such branches as geometry, number theory, calculus, differential equations, real and complex analysis, countable algebra, the topology of complete separable metric spaces, mathematical logic, computability theory). The emphasis is laid on Hilbert's program and the emerging reverse mathematics. The latter attempts to find weakest subsystems of second-order arithmetic Z_2 in which particular results of ordinary mathematics can be proved (this is done by deriving axioms from theorems).

The book consists of an Introduction, two main parts and an Appendix. The extensive Introduction is devoted to the description of the main subsystems of second-order arithmetic studied in the book and of mathematics within them, as well as to the main ideas of reverse mathematics. Part A consists of 5 chapters. Each chapter is devoted to one subsystem of Z_2 and to the development of mathematics in it. Particular chapters focus on: RCA_0 (recursive comprehension), ACA_0 (arithmetical comprehension), WKL_0 (weak König's lemma), ATR_0 (arithmetical transfinite recursion) and $\Pi_1^1 - CA_0$ (Π_1^1 comprehension). In Part B, consisting of 3 chapters, models of Z_2 and its subsystems are studied. In particular, Chapter VII is devoted to β-models, Chapter VIII to ω-models and Chapter IX to non-ω-models. In the Appendix some additional reverse mathematics results and problems are presented (without proofs but with references to the published literature).

The book is supplemented by historical and bibliographical notes (spread throughout the text), the extensive bibliography (279 items) and an index. This monograph can be studied both by graduate students in mathematical logic and foundations of mathematics and by experts. It provides an encyclopedic treatment of subsystems of Z_2. It can be viewed as a continuation of "Grundlagen der Mathematik" by D. Hilbert and P. Bernays.

R.Murawski (Poznań)

Zbl 968.60093
Schramm, Oded
Scaling limits of loop-erased random walks and uniform spanning trees.
Isr. J. Math. 118, 221-288 (2000).

The author introduces and analyses the so-called stochastic Löwner evolution (SLE), also named Schramm's process by other authors, which is the conjectured scaling limit for at least two interesting models from statistical mechanics. The existence of a scaling limit for these two models has not been proven yet, but the existence of scaling limits along suited subsequences is easily verified. Under the assumption that the limit actually exists and is conformal invariant, the author identifies them in terms of the SLE and derives some almost sure properties.

182

Author profile:

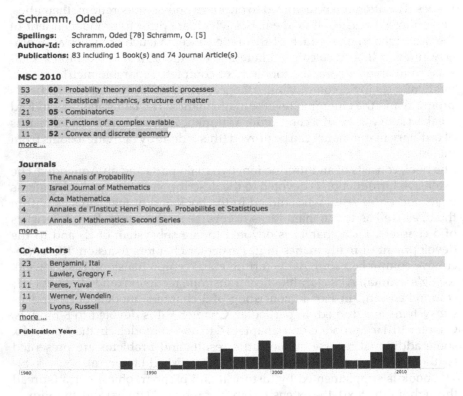

Schramm, Oded

Spellings: Schramm, Oded [78] Schramm, O. [5]
Author-Id: schramm.oded
Publications: 83 including 1 Book(s) and 74 Journal Article(s)

MSC 2010

53	**60** · Probability theory and stochastic processes
29	**82** · Statistical mechanics, structure of matter
21	**05** · Combinatorics
19	**30** · Functions of a complex variable
11	**52** · Convex and discrete geometry

more ...

Journals

9	The Annals of Probability
7	Israel Journal of Mathematics
6	Acta Mathematica
4	Annales de l'Institut Henri Poincaré. Probabilités et Statistiques
4	Annals of Mathematics. Second Series

more ...

Co-Authors

23	Benjamini, Itai
11	Lawler, Gregory F.
11	Peres, Yuval
11	Werner, Wendelin
9	Lyons, Russell

more ...

Publication Years

1980   1990   2000   2010

The notion of conformal invariance is around in the physicist's literature since a few decades, but has not been specified yet for many important models. The present paper gives a mathematically rigorous sense to the conformal invariance for two interesting models, the loop-erased random walk (LERW) and for the uniform spanning tree (UST). (The author announces to describe also the conjectured scaling limit of critical site percolation by similar means in a forthcoming paper.) Together with recent results on non-intersection exponents for Brownian motions, obtained by the author in collaboration with Lawler and Werner, these are the first examples of this kind and represent a breakthrough in the mathematical understanding of critical phenomena of two-dimensional models from statistical mechanics.

The LERW is a discrete-time random process on $\mathbb{Z}^d$ (here: $d = 2$) evolving in time where any loop that the trajectory closes is immediately removed, such that we obtain a self-avoiding random path. The UST is a random cycle-free connected subgraph of a given finite graph G that contains all the vertices and has the uniform distribution on the set of all such graphs. The notion of a UST may naturally be extended to infinite graphs G, and in the present paper the case of $G = \mathbb{Z}^2$ is considered throughout. There are intimate connections between the LERW and the UST. Two of the main open questions about these two models (and many related ones) are the following. Assume that the above models are defined on the lattice $\delta \mathbb{Z}^2$ rather than on $\mathbb{Z}^2$, is there a natural limiting random process for a properly scaled version of the above process as the mesh $\delta > 0$ tends to zero, and how can this limit be described? The present paper does not answer the first question, but (assuming the answer yes to the first one) the second.

The stochastic Löwner evolution is defined as follows. Let $(B(t))_{t\in[0,\infty)}$ be a standard Brownian motion on the boundary of the unit disc $\mathbb{U} = \{z \in \mathbb{C}: |z| < 1\}$, and fix a parameter $\kappa \geq 0$. With $\zeta(t) = B(-\kappa t)$ for $t \leq 0$, solve the so-called Löwner differential equation

$$\frac{\partial f}{\partial t} = z f_t'(z) \frac{\zeta(t) + z}{\zeta(t) - z}, \qquad z \in \mathbb{U}, \quad t \leq 0,$$

with the boundary value $f_0(z) = z$. Then f_t is a conformal mapping from $\mathbb{U}$ into some domain D_t. The process $(\mathbb{U} \setminus D_t)_{t\leq0}$ is called the SLE. Assuming that the scaling limit of the LERW exists and is conformal invariant, it is proven that, for the choice $\kappa = 2$, this scaling limit has the same distribution as $(f_t(\zeta(t)))_{t\leq0}$. An analogous assertion is proved for the UST. The choices $\kappa = 6$ and $\kappa = 8$ also lead to interesting processes in terms of which the conjectured scaling limit of critical site percolation and the Peano curve winding around the scaling limit of UST may be described in future work, respectively.   *Wolfgang König (Leipzig)*

Zbl 1077.11503
Borcherds, Richard E.
Problems in moonshine.
Yang, Lo (ed.) et al., First international congress of Chinese mathematicians. Proceedings of the ICCM '98, Beijing, China, December 12–16, 1998. Providence, RI: American Mathematical Society (AMS) (ISBN 0-8218-2652-2/pbk). AMS/IP Stud. Adv. Math. 20, 3-10 (2001).

The author presents open problems in Moonshine, the area in which he was awarded the Fields medal in 1998. After a short introduction describing the original Moonshine conjectures of McKay, Thompson, Conway, and Norton, some fifteen active problems in Moonshine are discussed individually. Here is a very brief enumeration of these problems: (1) Find a natural construction of the monster vertex algebra V. (2) Construct a good integral quadratic form on V. (3) Find an integral affine ind scheme for the monster Lie algebra. (4) Prove Norton's generalized Moonshine conjectures. (5) Explain any connection with the Y presentation of the Bimonster. (6) Classify related generalized Kac-Moody algebras. (7) Find all related hyperbolic reflection groups. (8) Find natural constructions for related generalized Kac-Moody algebras. (9) Explain any Moonshine for the non-Monstrous sporadic groups. (10) Explain McKay's relationship between the E_8, E_7, and E_6 Dynkin diagrams and the Monster, Baby Monster, and Fischer 24 sporadic groups. (11) Explain any

184

connection with mirror maps for $K3$ surfaces. (12) Find a monster manifold. (13) Explain any connection with complex hyperbolic reflection groups. (14) Find a nice moduli space related to $II_{1,25}$. (15) Explain any generalizations to L-functions or Dirichlet series of cusp forms. *Olaf Ninnemann (Berlin)*

Author profile:

Borcherds, Richard E.

Spellings: Borcherds, Richard E. [26] Borcherds, Richard [2] Borcherds, R.E. [2]
Author-Id: borcherds.richard-e
Publications: 30 including 23 Journal Article(s)

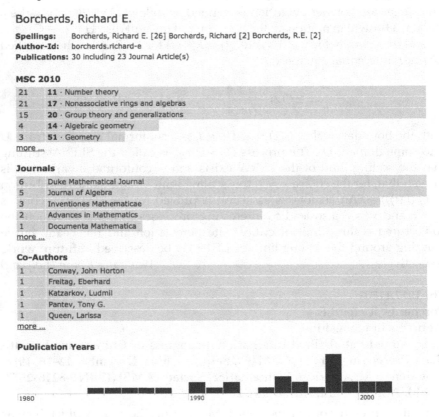

MSC 2010

21	11 · Number theory
21	17 · Nonassociative rings and algebras
15	20 · Group theory and generalizations
4	14 · Algebraic geometry
3	51 · Geometry

more ...

Journals

6	Duke Mathematical Journal
5	Journal of Algebra
3	Inventiones Mathematicae
2	Advances in Mathematics
1	Documenta Mathematica

more ...

Co-Authors

1	Conway, John Horton
1	Freitag, Eberhard
1	Katzarkov, Ludmil
1	Pantev, Tony G.
1	Queen, Larissa

more ...

Publication Years

1980   1990   2000

Zbl 1030.46005
Gowers, W.T.
An infinite Ramsey theorem and some Banach-space dichotomies.
Ann. Math. (2) 156, No.3, 797-833 (2002).

One of the main results of this deep paper (which circulated several years among specialists in preprint form) is the following dichotomy (Theorem 1.4): Every Banach space contains a subspace which either has an unconditional basis or is hereditarily indecomposable. This result is applied to solve an old problem of *S. Banach* who asked if the separable Hilbert space is the only homogeneous Banach space (that is an infinite-dimensional Banach space isomorphic to any of its infinite-dimensional subspaces). Combining the above dichotomy with the results of *W. Gowers* and *B. Maurey* [J. Am.

Math. Soc. 6, 851-874 (1993; Zbl 827.46008)] (asserting that a hereditarily indecomposable Banach space is isomorphic to no proper subspace of itself) and
of R. *Komorowski* and N. *Tomczak-Jaegermann* [Isr. J. Math. 89, 205-226 (1995;
Zbl 830.46008)] (asserting that each homogeneous Banach space either is isomorphic to a Hilbert space or else fails to have an uncounditional basis) the
author concludes that each homogeneous Banach space is isomorphic to the
separable Hilbert space, thus answering the Banach problem affirmatively.

Author profile:

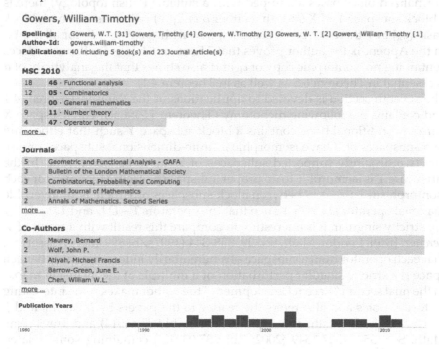

The mentioned dichotomy is a consequence of another Banach space dichotomy that has Ramsey-theoretic nature and by its spirit is close to the
classical Nash-Williams Theorem [Proc. Camb. Philos. Soc. 61, 33-39 (1965;
Zbl 129.00602)]. At first some notation. Given a Banach space X with a
fixed basis let Σ_f be the set of all finite block bases, that is sequences
$x_1 < \cdots < x_n$ of non-zero vectors from the unit ball of X, where $x < y$
means that $\max \operatorname{supp}(x) < \min \operatorname{supp}(y)$. A subspace Y of X generated by an
infinite block basis $x_1 < x_2 < \ldots$ is called a block subspace of X. The second
Banach space dichotomy (Theorem 2.1 proved in the second and third sec-
tion) asserts that given a Banach space X with a fixed basis, a family $\sigma \subset \Sigma_f$,
and an infinite sequence $\Delta = (\delta_n)$ of positive reals, one can find a block subspace Y of X such that either no sequence $y_1 < \cdots < y_n$ of vectors of Y belongs
to σ or else the second player has a winning strategy in the following infinite
two-person game $\sigma_\Delta[Y]$: At the n-th inning the first player selects an infinite-
dimensional block subspace Y_n of Y while the second player responds with

a vector $y_n > y_{n-1}$ in Y_n. The second player wins the game $\sigma_\Delta[Y]$ if for some n there is a sequence $x_1 < \cdots < x_n$ in σ such that $\|y_i - x_i\| \leq \delta_i$ for all $i \leq n$. The main result of the fourth and fifth section is Theorem 4.1, which is an infinite version of Theorem 2.1 (the principal difference is that Σ_f is replaced by a collection Σ of block bases, and σ is an analytic subspace of Σ with respect to a suitable Polish topology on Σ). In section 6 for a Banach space $X \supset c_0$, Theorem 4.1 is strengthened as follows (Theorem 6.5): For any sequence $\Delta = (\delta_n)$ of positive reals and any analytic subset σ of the space $\Sigma_1(X)$ of normalized block bases (equipped with a suitable Polish topology) there is a block subspace Y of X such that either $\sigma \cap \Sigma_1(Y) = \emptyset$ or else for each block basis (y_n) in Y there is a block basis (x_n) in σ with $\|y_i - x_i\| \leq \delta_n$ for all $n \in \mathbb{N}$. In the Appendix the author proves that Theorem 6.5 fails for Banach spaces containing no isomorphic copy of c_0 and also shows that the analyticity of σ is essential in Theorem 6.5 as well as in Theorem 4.1.

The seventh section is devoted to applications of the Ramsey Theorem 4.1 and contains the following dichotomy (Theorem 7.2): Each Banach space X with unconditional basis contains a block subspace Y such that either any two subspaces of Y have isomorphic infinite-dimensional subspaces or else any two disjointly supported subspaces of Y fail to be isomorphic. In the latter case Y is isomorphic to no proper subspace of itself. Moreover, for each isomorphism $T : Z \to W$ between block subspaces of Y there is an invertible diagonal operator $D : Y \to Y$ such that the operators $T - D\,|_Z$ and $T^{-1} - D^{-1}|_Z$ are strictly singular. It is interesting to compare this result with a result of V. Ferenczi [Bull. London Math. Soc 29, 338-344 (1997; Zbl 899.46010)] asserting that each operator from a subspace of a hereditarily indecomposable Banach space is a strictly singular perturbation of a multiple of the inclusion map.

In the final section "Recent Development" the author makes some interesting historical notes and also refers the reader to the papers by J. Bagaria and J. López-Abad [Adv. Math. 160, 133-174 (2001; Zbl 987.46014)] and Trans. Am. Math. Soc. 354, 1327-1349 (2002; Zbl 987.03040)] containing some related results and generalizations. *Taras Banakh (Lviv)*

Zbl 1130.53001

Perelman, Grisha

The entropy formula for the Ricci flow and its geometric applications.

arXiv e-print service, Cornell University Library, Paper No. 0211159, 39 p., electronic only (2002)

This is the first part of a masterpiece of mathematics which leads to a proof of the Poincaré and geometrization conjectures. Although the most striking application is in dimension 3 a good deal of the results presented here are valid in an arbitrary dimension n. This is the case for sections 1 to 10 of the present paper. It concerns the Ricci flow introduced by Richard Hamilton in the celebrated paper [R. Hamilton, J. Differ. Geom. 17, 255–306 (1982; Zbl 504.53034)]. Brief descriptions of each section follow. The detailed proofs can be read in [B. Kleiner and J. Lott, "Notes on Perelman's papers", arXiv:

Author profile:

Perelman, Grisha

Spellings: Perelman, Grisha [9] Perelman, G. [2] Perel'man, G. [1]
Author-Id: perelman.grisha
Publications: 12 including 8 Journal Article(s)

MSC 2010

12	**53** · Differential geometry
3	**57** · Manifolds and cell complexes
1	**58** · Global analysis, analysis on manifolds

Journals

3	Separata
1	Geometric and Functional Analysis - GAFA
1	Journal of Differential Geometry
1	Journal of the American Mathematical Society
1	Mathematische Zeitschrift

more ...

Co-Authors

1	Burago, Yury Dmitri (ed.)
1	Gromov, Mikhael

Publication Years

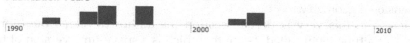

1990　　　　　　　　　　2000　　　　　　　　　2010

math.DG/0605667 (2006)], [*J. Morgan* and *G. Tian*, Ricci flow and the Poincaré conjecture, Clay Mathematics Monographs 3 (2007; Zbl 1179.57045)] and [*H.-D. Cao* and *X.-P. Zhu*, Asian J. Math. 10, No. 2, 165–492 (2006; Zbl 1200.53057)]. Section 1: On a closed manifold M the Ricci flow is presented as a gradient flow of a functional defined on the couples (g, f), where g is a Riemannian metric and f is a function on M. the function is chosen so that $e^{-f} d\,\mathrm{vol}_g$ is a fixed measure. The functional is

$$\mathcal{F}(g, f) = \int_M (R + |\nabla f|^2)e^{-f} d\,\mathrm{vol}_g,$$

in which R denotes the scalar curvature of the metric g and the norm is taken for this metric.

Section 2: This section and the following one present immediate corollaries which we can summarize as follows: On the space obtained by moding out the space of metrics by the action of the diffeomorphisms group and the homotheties there are no closed trajectories of the Ricci flow. In Perelman's terminology one shows that there are no non trivial breathers. This section is devoted to the proof of this assertion in the case of steady and expanding breathers. This relies on introducing very interesting invariants which are nondecreasing along the flow and stationary on Ricci solitons. One is the infinum on f of $\mathcal{F}(g, f)$ which turns out to be the smallest eigenvalue of

$-4\Delta + R$. The other one is the same eigenvalue normalised by the volume raised at the suitable power in order to get a scale invariant quantity.

Section 3: This section deals with the shrinking breathers which are more difficult to study. A modification of $\mathcal{F}$ called $\mathcal{W}$ is introduced. It is a function of three parameters: a metric g, a function f and a number τ which is $t_0 - t$ where t is the parameter of the Ricci flow. Again $\mathcal{W}$ is nondecreasing along the Ricci flow when f is normalized as in section 1 and is stationary on gradient shrinking solitons.

Section 4: This is the second breakthrough of this paper. It presents a tool that gives its full strength to the Ricci flow, namely a solution on a closed manifold M is non collapsed at a finite time t. It means there exists a number $\kappa > 0$ such that balls at time t and of radius $r \le \rho$ on which the Riemann curvature is bounded by r^{-2} have volume bounded below by κr^n. The number , called the scale, is the value below which this property is true. All these quantities are scaled invariant which makes this property true even after rescaling the metrics. This implies a local injectivity radius bound and allows to apply compactness theorems. This is crucial for the singularities analysis.

Section 5 and 6: These two sections are less useful at the moment. They intend to give justifications and explanations for the previous construction either using a statistical mechanics' approach or the point of view of infinite dimensional geometry.

Section 7: In this very important section the author develops the Morse theory of a functional called $\mathcal{L}$-length, which is a space-time version of the standard length (or energy). $\mathcal{L}$-geodesics, $\mathcal{L}$-Jacobi fields and $\mathcal{L}$-exponential are computed leading to some comparison theorems. The reduced volume is introduced, a monotonic quantity along the Ricci flow which allows to give an alternative proof of the nonlocal collapsing property. This is a key section for the three dimensional applications of the Ricci flow.

Section 8: A refined version of the nonlocal collapsing theorem is proved using the tools developed in the previous section, in particular the reduced volume.

Section 9: A differential Harnack inequality is proved.

Section 10: In this section it is proved that despite the fact that the Ricci flow is not local, it is pseudo-local. Namely, if the curvature is close to zero in a region (and with extra assumption), at least for a short time it remains not too far from zero, i.e., for a short time it is not affected by the possible presence of big curvature elsewhere on the manifold. This fact is made precise.

Section 11: This is a section devoted to the classification of the so-called κ-solutions in dimension 3. They are solutions of the Ricci flow which are ancient (i.e., have an infinite past), non flat, have bounded and nonnegative curvature operator on each time slice and are κ-non-collapsed at all scales for a positive κ. They are infinitesimal models for the singularities of a three-dimensional Ricci flow. Indeed, in three dimensions, they are obtained as limits of suitable blow-up around a singularity (points of very large curvature) and at finite time. This classification is necessary in order to understand the Ricci flow with surgery.

Section 12: This section presents another breakthrough, the so-called canonical neighbourhood theorem. This is again specific to the three-dimensional case. It is shown that for a smooth Ricci flow, with normalised initial data, there is a universal number such that if at some point the scalar curvature becomes larger than this value, then a neighbourhood of this point, of controlled size, is, after rescaling, close to a piece of a κ-solution. This describes quite precisely the geometry of the manifold, whose evolution is given by the Ricci flow equation, around points of large curvature. The rest of the section is devoted to the (beginning of the) long term analysis of the Ricci flow.

Section 13: In this section the proof of geometrisation is sketched for manifolds which carry a Ricci flow defined for all time. It is shown that the manifold has a thick-thin decomposition, that the thick part becomes hyperbolic and is bounded by incompressible tori following arguments due to R. Hamilton. The thin part is claimed without proof to be a graph manifold. Some arguments which are adapted from the works of R. Hamilton are more precisely treated in [G. Perelman, "Ricci flow with surgery on three-manifolds", arXiv: math.DG/0303109 (2003; Zbl 1130.53002)]. Others concerning the case where the flow develops singularities are sketched but not justified. The reader is referred to [Perelman, loc. cit.].   Gérard Besson (Grenoble)

Zbl 1130.53002
Perelman, Grisha
Ricci flow with surgery on three-manifolds
arXiv e-print service, Cornell University Library, Paper No. 0303109, 22 p., electronic only (2003)

This is the second paper written by G. Perelman which proves the geometrization conjecture. It is presented by the author as a "technical paper" in which some assertions made in [G. Perelman, "The entropy formula for the Ricci flow and its geometric applications", arXiv: math.DG/0211159 (2002; Zbl 1130.53001)] are proved and some others are corrected. It presents the Ricci flow with surgery on three-manifolds which is a version of the Ricci flow taking into account the singularities. It is inspired by the construction made by R. Hamilton in [Commun. Anal. Geom. 5, No. 1, 1–92 (1997; Zbl 892.53018)]. As for G. Perelman (loc. cit.) the details can be found in [B. Kleiner and J. Lott, "Notes on Perelman's papers", arXiv: math.DG/0605667 (2006), J. Morgan and G. Tian, Ricci flow and the Poincaré conjecture, Clay Mathematics Monographs 3 (2007; Zbl 1179.57045) and H.-D. Cao and X.-P. Zhu, Asian J. Math. 10, No. 2, 165–492 (2006; Zbl 1200.53057)].

Sections 1 and 2: These sections present preliminary material on ancient and standard solutions. In section 1 the author gives a more precise classification of ancient solutions by the study of their asymptotic soliton. Let us recall that the ancient solutions are those which have an infinite past and ought to be (with some more properties) the infinitesimal models for the singularities. The asymptotic soliton is a limit when time goes to $-\infty$ of a rescaling of the ancient solution. These notions are defined in section 11 of Perelman (loc. cit.) and more refined properties are described here.

Section 2 is devoted to the standard solution. This is a solution of the Ricci flow whose initial data is a half cylinder capped-off by a hemisphere. The initial data being non compact existence and uniqueness of the solution is not immediate. It is shown that the Ricci flow with such an initial data exists in the time interval $[0, 1)$ and is unique. An extremely detailed proof is given in [Morgan and Tian (loc. cit.), see also Kleiner and Lott (loc. cit.) and Cao and Zhu (loc. cit.)].

Section 3: In this important section the first time for which the flow becomes singular is described. Being singular means that the scalar curvature becomes infinite somewhere. If it is infinite everywhere then the manifold is covered by canonical neighbourhoods and its topology is thus completely known. If it becomes infinite on a subset Ω, it is open and one can describe the part of Ω of high scalar curvature. They are, again, covered with canonical neighbourhoods. This leads to the important notion of horns which are subsets of Ω diffeomorphic to open half cylinders and on which the scalar curvature goes to infinity on one end. It is in these horns that the surgery will take place. The subset $M \setminus \Omega$ is covered by canonical neighbourhoods and is topologically simple (and known).

Sections 4 and 5: These are the sections in which the Ricci flow with surgery is defined. In section 4 the so-called Ricci flow with cutoff appears. It is given by a collection of smooth Ricci flows defined on a 3-manifold on adjacent intervals of time. On the boundary of these intervals surgeries take place. In this section it is supposed that the flow satisfies the a priori assumptions which are: the Hamilton-Ivey pinching property and the canonical neighbourhood property for points whose scalar curvature is larger than a parameter $r > 0$. The surgery is done in horns. More precisely it is shown that if one goes "far enough" into a horn then one finds a neck of size $2/\delta$ for some parameter $\delta > 0$ small enough and of curvature close to $h > 0$ another parameter depending on r and δ. The flow is called a Ricci flow with δ-cutoff. Now, this δ-neck is cut in the middle and a suitably rescaled compact piece of the standard solution is glued on one side. The flow may start again with the new Riemannian manifold thus obtained. An important property is the fact that the added piece (called an almost-standard cap) remains close to the evoluting standard solution for a while.

In section 5 the author shows that the flow can be defined, satisfying the a priori assumptions for all time. The parameters r and δ must now depend on the time parameter and one issue is to prove that they do not go to zero in finite time. The proof is very close to the proof of the canonical neighbourhood theorem done in [Perelman (loc. cit.)] section 12, taking into account the surgeries. This is the key result of this series of work by G. Perelman.

Sections 6 and 7: Now, the long time behaviour of the Ricci flow with surgery is analysed carefully. Section 6 presents technical issues such as curvature estimates in the future and the past of a given time slice.

Section 7 defines and describes the thick-thin decomposition of the manifold. It is inspired by R. Hamilton [Commun. Anal. Geom. 7, No. 4, 695–729 (1999; Zbl 939.53024)]. It is shown that thick parts become more and more hyperbolic bounded by incompressible tori. The description of the thin part relies on an unpublished paper by the author; the conclusion is that it is a graph manifold.

Section 8: It contains an alternative approach. The geometric decomposition is described using the values of a Riemannian invariant as threshold for the different possibilities. This invariant is the first eigenvalue of a Schrödinger operator whose potential is given by the scalar curvature. A simpler argument can be found in [Kleiner and Lott (loc. cit.), section 93].

At the time when this review is written the fact that the thin part is a graph manifold is still a bit controversial although it is widely believed to be true. Some details are missing in the literature. An alternative approach is given in [L. Bessières, G. Besson, M. Boileau, S. Maillot and J. Porti, "Suites de métriques extraites du flot de Ricci sur les variétés asphériques de dimension 3", arXiv:0706.2065]. *Gérard Besson (Grenoble)*

Zbl 1130.53003
Perelman, Grisha
Finite extinction time for the solutions to the Ricci flow on certain three-manifolds.
arXiv e-print service, Cornell University Library, Paper No. 0307245, 7 p., electronic only (2003)

In this short paper, posted in july 2003, G. Perelman shows that for any closed oriented three-manifold M, whose prime decomposition contains no aspherical factors, the Ricci flow with surgery stops in finite time. To stop means that the scalar curvature becomes large everywhere on M so that it is covered by canonical neighbourhoods and hence the topology is completely known; it is also said that the flow becomes extinct in finite time. It is an important step in the proof of the Poincaré conjecture; indeed it is a short cut which avoids using the long time behaviour of the Ricci flow described in [G. Perelman, "Ricci flow with surgery on three-manifolds", arXiv: math.DG/0303109 (2003; Zbl 1130.53002)] sections 6 to 8.

The idea is to fill some suitably chosen loop by a minimal disc and let both the metric evolve by the Ricci flow and the loop by the curve shortening flow. This idea is similar to one used by R. Hamilton in [R. Hamilton, Commun. Anal. Geom. 7, No. 4, 695–729 (1999; Zbl 939.53024)] and a very detailed proof is given in [J. Morgan and G. Tian, Ricci flow and the Poincaré conjecture, Clay Mathematics Monographs 3 (2007; Zbl 1179.57045)]. An alternative approach using harmonic maps is given in [T. Colding and W. Minicozzi, J. Am. Math. Soc. 18, No. 3, 561–569 (2005; Zbl 1083.53058)] and with more details in [T.

Colding and *W. Minicozzi*, "Width and finite extinction time of Ricci flow",
arXiv:0707.0108].   *Gérard Besson (Grenoble)*

Zbl 1130.11039
Kleinbock, Dmitry; Lindenstrauss, Elon; Weiss, Barak
On fractal measures and Diophantine approximation.
Sel. Math., New Ser. 10, No. 4, 479-523 (2004).

Author profile:

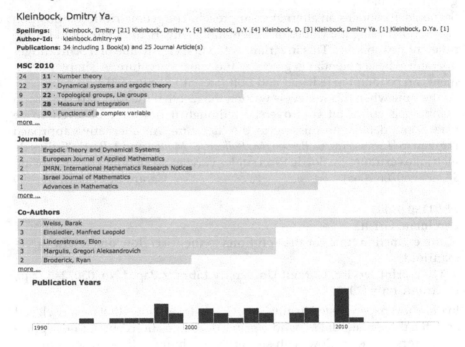

Kleinbock, Dmitry Ya.

Spellings: Kleinbock, Dmitry [21] Kleinbock, Dmitry Y. [4] Kleinbock, D.Y. [4] Kleinbock, D. [3] Kleinbock, Dmitry Ya. [1] Kleinbock, D.Ya. [1]
Author-Id: kleinbock.dmitry-ya
Publications: 34 including 1 Book(s) and 25 Journal Article(s)

MSC 2010

24	11 · Number theory
22	37 · Dynamical systems and ergodic theory
9	22 · Topological groups, Lie groups
5	28 · Measure and integration
3	30 · Functions of a complex variable

more ...

Journals

2	Ergodic Theory and Dynamical Systems
2	European Journal of Applied Mathematics
2	IMRN. International Mathematics Research Notices
2	Israel Journal of Mathematics
1	Advances in Mathematics

more ...

Co-Authors

7	Weiss, Barak
3	Einsiedler, Manfred Leopold
3	Lindenstrauss, Elon
3	Margulis, Gregori Aleksandrovich
2	Broderick, Ryan

more ...

Publication Years

1990   2000   2010

The paper under review is concerned with an extension of Mahler's problem
in metric number theory, originally resolved by *V. G. Sprindzuk* ["Mahler's
problem in metric number theory," Nauka i Teknika, Minsk (1967; Zbl
168.29504)]. Mahler conjectured that almost no points with respect to the
natural measure on the curve $(x, x^2, \ldots, x^n)$ should be no better approximable
by rationals with the same denominator than is the case for generic points in
$\mathbb{R}^n$. Here the approximation is measured in the sup-norm. A subset E of $\mathbb{R}^n$
supporting a natural measure is said to be *extremal* if it satisfies this property.
If we instead consider multiplicative approximation, where the distances of
the individual coordinates are multiplied, we arrive at the notion of *strong
extremality*, which implies extremality.
In a far-reaching extension of Mahler's conjecture, *D. Y. Kleinbock* and *G. A.
Margulis* [Ann. Math. (2) 148, No.1, 339–360 (1998; Zbl 922.11061)] showed
that manifolds satisfying a certain non-degeneracy condition are strongly
extremal.

Author profiles:

Lindenstrauss, Elon

Spellings: Lindenstrauss, Elon [38]
Author-Id: lindenstrauss.elon
Publications: 38 including 1 Book(s) and 32 Journal Article(s)

MSC 2010

31	37 · Dynamical systems and ergodic theory
15	11 · Number theory
11	22 · Topological groups, Lie groups
11	28 · Measure and integration
4	58 · Global analysis, analysis on manifolds

more ...

Journals

5	Israel Journal of Mathematics
4	Annals of Mathematics. Second Series
3	Ergodic Theory and Dynamical Systems
3	IMRN. International Mathematics Research Notices
2	Comptes Rendus. Mathématique. Académie des Sciences, Paris

more ...

Co-Authors

9	Einsiedler, Manfred Leopold
4	Bourgain, Jean
4	Venkatesh, Akshay
3	Kleinbock, Dmitry Ya.
3	Michel, Philippe

more ...

Publication Years

1990　　　　　　　　2000　　　　　　　　2010

Weiss, Barak

Spellings: Weiss, B. [44] Weiss, Barak [24] Weiß, B. [2]
Author-Id: weiss.barak
Publications: 70 including 1 Book(s) and 61 Journal Article(s)

MSC 2010

28	37 · Dynamical systems and ergodic theory
18	28 · Measure and integration
13	22 · Topological groups, Lie groups
12	11 · Number theory
10	54 · General topology

more ...

Journals

9	Ergodic Theory and Dynamical Systems
8	Israel Journal of Mathematics
5	Proceedings of the American Mathematical Society
4	Geometric and Functional Analysis - GAFA
2	Duke Mathematical Journal

more ...

Co-Authors

10	Glasner, Eli
7	Kleinbock, Dmitry Ya.
5	Smillie, John
3	Konheim, Allan G.
3	Thouvenot, Jean-Paul

more ...

Publication Years

1960　　　　1970　　　　1980　　　　1990　　　　2000　　　　2010

The present paper extends and resolves Mahler's conjecture further. We say that a measure is (strongly) extremal if the support of the measure satisfies the (strong) extremality property. A measure μ on $\mathbb{R}^n$ is said to be *friendly* if it is doubling (or Federer) almost everywhere, if the μ-measure of any affine hyperplane is zero, and if the measure satisfies a certain technical decay condition. The main result of the paper under review is that friendly measures are strongly extremal.

The class of friendly measures include the volume measure on non-degenerate manifolds, and so the result of *Kleinbock* and *Margulis* [loc.cit.] is contained in the main result. In fact, this is part of a larger class of push-forwards of the so-called absolutely friendly measures. Other examples are also given. The Hausdorff s-measure restricted to an s-dimensional attractor of an iterated function system of affine contractions satisfying the open set condition is friendly. The same holds for direct products of friendly measures.

The main theorem is proved by extending the method of *Kleinbock* and *Margulis* [loc.cit.], re-interpreting the strong extremality property as a certain quantitative non-divergence property for a certain flow in the homogeneous space $\mathrm{SL}(n,\mathbb{R})/\mathrm{SL}(n,\mathbb{Z})$. It is then shown that the above measures are all friendly. The paper is concluded with a section on related problems, results and conjectures. *Simon Kristensen (Århus)*

Zbl 1132.11034
McMullen, Curtis T.
Minkowski's conjecture, well-rounded lattices and topological dimension.
J. Am. Math. Soc. 18, No. 3, 711-734 (2005).
One of the great problems of the geometry of numbers is the following conjecture on the product of non-homogeneous linear forms: For $x \in \mathbb{R}^n$ let $|x|^2 = x_1^2 + \cdots + x_n^2$ and $N(x) = |x_1 \cdots x_n|$. Then for any lattice L in $\mathbb{R}^n$ with determinant 1 holds
$$\sup_{x \in \mathbb{R}^n} \inf_{y \in L} N(x - y) \leq 2^{-n}.$$

Equality holds if and only if $L = D\mathbb{Z}^n$ where D is a diagonal matrix with positive entries and determinant 1. The conjecture has been proved for $n = 2, 3, 4, 5$ by [*M. Minkowski*, Diophantische Approximation. Neudruck. Würzburg: Physica-Verlag (1961; Zbl 103.03403)], *R. Remak* [Math. Z. 17, 1–34, 18, 173–200 (1923; JFM 49.0101.03)], *F. J. Dyson* [Ann. Math. (2) 49, 82–109 (1948; Zbl 31.15402)] and *B. F. Skubenko* [Zap. Nauchn. Semin. Leningr. Otd. Mat. Inst. Steklova 33, 6–36 (1973; Zbl 352.10010)]. For general n there are estimates due to *N. G. Chebotarev* [Vierteljahresschr. Naturforsch. Ges. Zürich 85, Beibl. 32, 27–30 (1940; Zbl 23.20701)] and others including *E. Bombieri* [Acta Math. 8, 273–281 (1963; Zbl 123.04505)]. For large sets of lattices the conjecture has been proved by *A. M. Macbeath* [Proc. Glasg. Math. Assoc. 5, 86–89 (1961; Zbl 98.26303)] and the reviewer [Acta Arith. 13, 9–27 (1967; Zbl 153.07102)]. One line of attack is to show the following:

(i) For each lattice L in $\mathbb{R}^n$ there is a diagonal matrix D as above such that DL has the following property: the set of vectors $y \in L\backslash\{o\}$ of minimum length $|y|$ span $\mathbb{R}^n$, i.e. DL is well rounded.

(ii) If L is well rounded and has determinant 1, then its covering radius satisfies

$$\sup_{x\in\mathbb{R}^n} \inf_{y\in L} |x - y| \le \frac{\sqrt{n}}{2}$$

where equality holds only in a particular case.

Author profile:

McMullen, Curtis T.

Spellings: McMullen, Curtis T. [46] McMullen, Curt [14] McMullen, C. [2] McMullen, C.T. [1]
Author-Id: mcmullen.curtis-t
Publications: 63 including 3 Book(s) and 55 Journal Article(s)

MSC 2010

37	**30** · Functions of a complex variable
35	**37** · Dynamical systems and ergodic theory
18	**32** · Functions of several complex variables and analytic spaces
13	**11** · Number theory
11	**57** · Manifolds and cell complexes
more ...	

Journals

7	Inventiones Mathematicae
5	Duke Mathematical Journal
4	Annals of Mathematics. Second Series
4	Bulletin (New Series) of the American Mathematical Society
3	Acta Mathematica
more ...	

Co-Authors

2	Elkies, Noam D.
2	Gross, Benedict H.
1	Brayton, Robert K.
1	Brooks, Robert
1	DeMarco, Laura G.
more ...	

Publication Years

If (i) and (ii) hold, the inequality of the arithmetic and geometric mean yields the conjecture. For more information see the reviewer and C. G. Lekkerkerker [Geometry of numbers. 2nd ed., North-Holland Mathematical Library, Vol. 37. Amsterdam etc.: North-Holland (1987; Zbl 611.10017)] and the reviewer [Convex and discrete geometry. Grundlehren der Mathematischen Wissenschaften 336. Berlin: Springer (2007; Zbl 1139.52001)].

Proposition (ii) has been proved by A. C. Woods [J. Number Theory 4, 157–180 (1972; Zbl 232.10020)] for $n = 6$. The author shows (i) for all lattices of determinant 1 with $N(L) > 0$ in all dimensions. Using a result of B. J. Birch and H. P. F. Swinnerton-Dyer [Mathematika Lond. 3, 25–39 (1956; Zbl 74.03702)] this finally yields the conjecture for $n = 6$. This result is an important contribution to the geometry of numbers.   *Peter M. Gruber (Wien)*

196

Zbl 1186.52011
Hales, Thomas C.
Sphere packings. III: Extremal cases.
Discrete Comput. Geom. 36, No. 1, 71-110 (2006).

Author profile:

Hales, Thomas C.

Spellings: Hales, Thomas C. [38] Hales, T.C. [3] Hales, Thomas [1]
Author-Id: hales.thomas-c
Publications: 42 including 29 Journal Article(s)

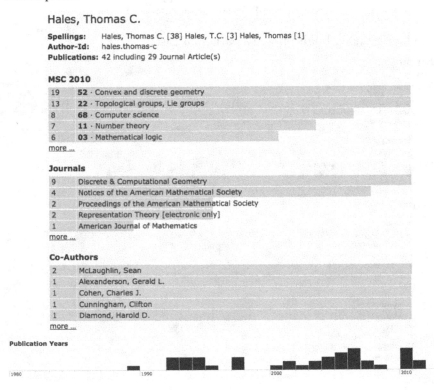

MSC 2010

19	**52** · Convex and discrete geometry
13	**22** · Topological groups, Lie groups
8	**68** · Computer science
7	**11** · Number theory
6	**03** · Mathematical logic

more ...

Journals

9	Discrete & Computational Geometry
4	Notices of the American Mathematical Society
2	Proceedings of the American Mathematical Society
2	Representation Theory [electronic only]
1	American Journal of Mathematics

more ...

Co-Authors

2	McLaughlin, Sean
1	Alexanderson, Gerald L.
1	Cohen, Charles J.
1	Cunningham, Clifton
1	Diamond, Harold D.

more ...

Publication Years

1980　　　　1990　　　　2000　　　　2010

This paper is the third in the series of six papers by the author devoted to the proof of the Kepler's conjecture, all in Discrete Comput. Geom. 36, no. 1. This famous conjecture, open since 1611, asserts that no packing of congruent balls in three dimensions has density greater than the face-centered cubic packing. In the previous paper of this series a compact topological space of decomposition stars was defined, and a continuous scoring function on this space was introduced. Moreover, Kepler's conjecture was related to a certain conjecture about location of global maxima of this scoring function.

The goal of the current paper is to prove that the two conjectured global maxima of the scoring function (corresponding to decomposition stars of the face-centered cubic and hexagonal-close packings) are at least local maxima. In addition, general upper bounds for the scoring function on various other decomposition stars are obtained.　　　　*Lenny Fukshansky (Claremont)*

Zbl 1186.52012
Hales, Thomas C.
Sphere packings. IV: Detailed bounds.
Discrete Comput. Geom. 36, No. 1, 111-166 (2006).

This paper is the fourth in the series of six papers by the author devoted to the proof of the Kepler's conjecture, all in Discrete Comput. Geom. 36, no. 1. This famous conjecture, open since 1611, asserts that no packing of congruent balls in three dimensions has density greater than the face-centered cubic packing. In the second paper of this series a compact topological space of decomposition stars was defined, and a continuous scoring function on this space was introduced. Moreover, Kepler's conjecture was related to a certain conjecture about location of global maxima of this scoring function. These conjectured points were shown to be at least local maxima in the third paper of this series.

The current paper contains the technical heart of the proof of Kepler's conjecture; its results rely on long computer calculations. The scoring function can be expressed as the sum of terms over different regions of the unit sphere. The main objective of this paper is to obtain good bounds on the values of the scoring function over such regions.   *Lenny Fukshansky (Claremont)*

Zbl 1186.52009
Ferguson, Samuel P.
Sphere packings. V: Pentahedral prisms.
Discrete Comput. Geom. 36, No. 1, 167-204 (2006).

This paper is the fifth in the series of six papers devoted to the proof of the Kepler's conjecture, all in Discrete Comput. Geom. 36, No. 1. This famous conjecture, open since 1611, asserts that no packing of congruent balls in three dimensions has density greater than the face-centered cubic packing. In the second paper of this series a compact topological space of decomposition stars was defined, and a continuous scoring function on this space was introduced. Moreover, Kepler's conjecture was related to a certain conjecture about location of global maxima of this scoring function. The two conjectured global maxima of the scoring function (corresponding to decomposition stars of the face-centered cubic and hexagonal-close packings) were shown to be local maxima in the third paper of this series.

A so-called *contravening* decomposition star is a potential counterexample to the Kepler's conjecture. This paper considers a particular class of potentially contravening decomposition stars, *pentahedral prisms*, on which the scoring function comes very close to its values at the decomposition stars of the face-centered cubic and hexagonal-close packings. The subject of this paper is to prove that in fact pentahedral prisms are not contravening.

Lenny Fukshansky (Claremont)

198

Author profile:

Ferguson, Samuel P.

Spellings: Ferguson, Samuel P. [2] Ferguson, S. [1]
Author-Id: ferguson.samuel-p
Publications: 3 including 3 Journal Article(s)

MSC 2010

2	52 · Convex and discrete geometry
1	86 · Geophysics

Journals

2	Discrete & Computational Geometry
1	Journal of Geodesy

Co-Authors

1	Hales, Thomas C.
1	Hammada, Y.
1	Heck, Bonnie S.
1	Kern, Martin
1	Novák, Petr

more ...

Publication Years

2000		2010

Zbl 1186.52013
Hales, Thomas C.
Sphere packings. VI: Tame graphs and linear programs.
Discrete Comput. Geom. 36, No. 1, 205-265 (2006).

This paper is the sixth and final in the series of papers devoted to the proof of the Kepler's conjecture, all in Discrete Comput. Geom. 36, No. 1. This famous conjecture, open since 1611, asserts that no packing of congruent balls in three dimensions has density greater than the face-centered cubic packing. In the second paper of this series a compact topological space of decomposition stars was defined, and a continuous scoring function on this space was introduced. Moreover, Kepler's conjecture was related to a certain conjecture about location of global maxima of this scoring function. The two conjectured global maxima of the scoring function (corresponding to decomposition stars of the face-centered cubic and hexagonal-close packings) were shown to be local maxima in the third paper of this series.

In the current paper the set of all points in the space of decomposition stars at which the scoring function assumes the conjectured maximal value is considered. A certain planar graph is then associated to each such point, and it is proved that each such graph is isomorphic to a *tame* graph. Now there are only finitely many isomorphism classes of tame graphs, and the linear programming methods are used to eliminate all possibilities except for the graphs corresponding to face-centered cubic, the hexagonal-close packings, and the pentahedral prisms decomposition stars. Since the penta-hedral prisms were eliminated in the previous paper of this series, Kepler's conjecture follows. The results of this paper rely on long computer calcula-tions. *Lenny Fukshansky (Claremont)*

Zbl 1129.01018
Scharlau, Winfried
Who is Alexander Grothendieck? Anarchy, mathematics, spirituality.
A biography. Part 1: Anarchy.
Havixbeck: Scharlau. 180 p. EUR 14.00 (2007).

Alexander Grothendieck, born March 28, 1928 in Berlin, Germany, is a math-ematician of French citizenship, who unanimously is considered to be one of the greatest minds in contemporary mathematics as a whole. His influ-ence on the development of several branches of modern pure mathematics in the second half of the 20th century was absolutely pioneering and pro-pelling, especially with regard to algebraic geometry, homological algebra, topology, arithmetic geometry, and functional analysis. He was awarded the Fields Medal in 1966 for his utmost fundamental contributions to the rigor-ous, abstract and sweeping reformation of algebraic geometry by developing completely new conceptual frameworks, methods, and tools of most pow-erful impact. In particular, Alexander Grothendieck became famous for his great mastery of highly abstract approaches to classical mathematics, for his unparalleled perfectionism in matters of systematizing and presenting the-ories, and for demonstrating how to derive new concrete results from very general methods.

On the other hand, apart from the unique mathematical genius Alexander Grothendieck, there is the just as unique personality of this outstanding man as a contemporary, socially and politically acting human being, colleague, teacher, next of kin, friend, and adversary, which is the subject of many haunt-ing stories and some misleading rumors about him. The book under review, written by a leading German mathematician and established novel writer, is the first part of a biographical trilogy about the uniquely fascinating and dramatic life of the "phenomenon" Alexander Grothendieck. Based upon extensive investigations, consultations of acquaintances of Grothendieck's, interpretations of some of his correspondences, and analyzing his 1000-page autobiographical manuscript "Récoltes et semailles" (1986), W. Scharlau has undertaken the rewarding attempt to present an authentic, objective and fair report on the first third of Alexander Grothendieck's exceptional life, thereby trying to illuminate the essential facts that formed his individual characteristic, philosophy of life, and social behaviour later on.

Accordingly, the current first volume is concerned with his ancestry, his parents, his early family life, his fate as a displaced person during the upheavals of World War II, his school time, his studies in mathematics, and his early career as a mathematical researcher in France.

The first nine sections of the book are devoted to the life story of Alexander Grothendieck's Russian Jewish father, Alexander Shapiro (1890–1942), and his anarchist political activities which finally ended with his tragic death in Auschwitz. The following four sections give a first depiction of Alexander Grothendieck's German mother, Johanna (Hanka) Grothendieck (1900–1957), her parents, her youth, and her first husband, Johannes (Alf) Raddatz (1897–1958). Born as Alexander Raddatz in 1928 in Berlin, Alexander Grothendieck spent the first five years of his life, together with his mother and his half-sister, Frode (Maidi) Raddatz, in Berlin, before having been entrusted to his foster-parents, the Heydorn family in Hamburg-Blankenese, when his mother Hanka joint his father, Alexander Shapiro (alias Tanaroff) in their political activities in France and Spain. This is described in Sections 14–20, together with the known facts about the end of Grothendieck's father in Auschwitz, while the following sections compile various details about the joint life of Hanka and Alexander Grothendieck in diverse refugee camps in France between 1940 and 1944 (Sections 21–23).

The remaining sections give particulars of the complicated relations between Alexander Grothendieck and his mother after 1945, of Alexander's growing into a mathematician during his studies in Montpellier, and of Hanka Grothendieck's unpublished autobiographical novel "A Woman" ("Eine Frau", written in German).

The present first volume concludes the report on Alexander Grothendieck's first 25 years of life with a description of his extended mathematical studies in Paris and Nancy, his rise to a great researcher under the influence of L. Schwartz, H. Cartan, J. Dieudonné, J.-P. Serre, and others, and of the dramatic clashes between him and his mother Hanka until her death in 1957.

Actually, the biographies of his parents, foster-parents, siblings, and other related folks play a predominant role in this part of the trilogy, just as the various unusual circumstances in his early life do, but it is exactly these facts that help understand the shaping of Alexander Grothendieck's unique individuality during that crucial period of his growing up. The author portrays Grothendieck's early associates and their paths through life in a very detailed and gripping manner, and that with a masterly feeling for correct idiom, and he places the background of the young Alexander Grothendieck precisely into this web of almost bizarre marches of destiny. No doubt, this is an essential key to understand his later mode of life, his unusual habits and politics, his confrontations with colleagues and French authorities, his left-wing and pacifist political views, his early withdrawal from mathematics, his reasons for retirement, his later writings, and finally his disappearance from social life in 1991.

In the forthcoming two volumes, the author plans to tell about Alexander Grothendieck's creative work as active researcher in mathematics during the period from 1950 until 1970, when he virtually revolutionized large parts of pure mathematics, and about his withdrawal from both mathematics and the human society afterwards, his meditations, and his mysterious behaviour ever since.

Winfried Scharlau is completely right when he emphasizes that the nearly incredible path of life of Alexander Grothendieck should be regarded as a vivid part of our society in the 20th century and thereafter, thus deserving very much interest, analysis, and profound understanding. He has portrayed the first third of Grothendieck's life without the slightest trace of sensationalism, as it shows through in many rumors about it, but with a sympathetic understanding for an exceptional human being whose spirituality and moral philosophy seems to have reached the borderline of human nature.

Werner Kleinert (Berlin)

Author profile:

Scharlau, Winfried

Spellings: Scharlau, Winfried [39] Scharlau, W. [15]
Author-Id: scharlau.winfried
Publications: 54 including 14 Book(s) and 33 Journal Article(s)

MSC 2010

20	11 · Number theory
17	01 · History; biography
13	12 · Field theory and polynomials
8	16 · Associative rings and algebras
7	15 · Linear and multilinear algebra; matrix theory

more ...

Journals

5	Inventiones Mathematicae
5	Mathematische Zeitschrift
3	Mathematische Annalen
2	Archive for History of Exact Sciences
2	Historia Mathematica

more ...

Co-Authors

4	Knebusch, Manfred
3	Hirzebruch, Friedrich Ernst Peter
2	Krüskemper, Martin
2	Opolka, Hans
2	Quebbemann, Heinz-Georg

more ...

Publication Years

Zbl 1191.11025
Green, Ben; Tao, Terence
The primes contain arbitrarily long arithmetic progressions.
Ann. Math. (2) 167, No. 2, 481-547 (2008).

Author profile:

Green, Benjamin

Spellings: Green, Ben [35] Green, Benjamin [1]
Author-Id: green.benjamin
Publications: 36 including 29 Journal Article(s)

MSC 2010

34	11 · Number theory
5	05 · Combinatorics
5	42 · Fourier analysis
2	20 · Group theory and generalizations
2	37 · Dynamical systems and ergodic theory

more ...

Journals

4	Annals of Mathematics. Second Series
3	Bulletin of the London Mathematical Society
3	Geometric and Functional Analysis - GAFA
2	Combinatorics, Probability and Computing
2	Mathematical Proceedings of the Cambridge Philosophical Society

more ...

Co-Authors

12	Tao, Terence C.
5	Ruzsa, Imre Z.
2	Sanders, Tom
1	Konyagin, Sergei V.
1	Sisask, Olof

Publication Years

2000   2010

This paper needs little introduction: in 2004, the authors proved [cf. Ann. Math. (2) 171, No. 3, 1753–1850 (2010; Zbl 05712763)] that the primes contain arbitrarily long arithmetic progressions, a startling result considering that the previous state of the art had been an infinitude of four term arithmetic progressions in which three elements were prime and the fourth a product of at most two primes. As frequently happens when an old problem falls, the solution also precipitated a vast new theory of linear forms in the primes which looks like it will lead to a resolution of the Hardy-Littlewood conjecture for essentially all systems except those describing structures such as twin primes or the Goldbach conjecture. This will be a major achievement, and although the theory has moved on somewhat from this opening of the door, it is still very much worth reading.

The main idea of the paper is a transference principle allowing the authors to transfer results from vanilla structures to pseudo-random versions. They then use some estimates of Goldston and Yıldırım to show that in a suitable sense subsets of the primes behave pseudo-randomly, which allows them to transfer Szemerédi's theorem to subsets of the primes. A good introduction to this sphere of ideas may be found in the earlier paper of Green where this result is proved for three term progressions.   *Tom Sanders (Cambridge)*

Author profile:

Tao, Terence C.

Spellings: Tao, Terence [184] Tao, T. [22] Tao, Terry [3] Tao, Terrence [1] Tao, Terence C. S. [1] Tao, Terence C. [1]
Author-Id: tao.terence-c
Publications: 212 including 15 Book(s) and 178 Journal Article(s)

MSC 2010

70	**35** · Partial differential equations (PDE)
67	**42** · Fourier analysis
44	**11** · Number theory
29	**37** · Dynamical systems and ergodic theory
19	**05** · Combinatorics

more ...

Journals

12	Geometric and Functional Analysis - GAFA
10	Mathematical Research Letters
8	American Journal of Mathematics
8	Journal of the American Mathematical Society
7	Duke Mathematical Journal

more ...

Co-Authors

18	Keel, Markus
16	Colliander, James E.
16	Thiele, Christoph Martin
14	Takaoka, Hideo
12	Green, Benjamin

more ...

Publication Years

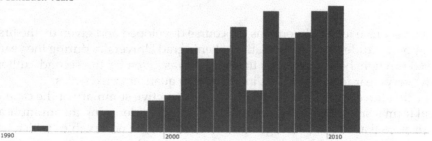

| 1990 | 2000 | 2010 |

Zbl 1193.81001
Faddeev, L. D.; Yakubovskiĭ, O. A.
Lectures on quantum mechanics for mathematics students.
Translated by Harold McFaden.
Student Mathematical Library 47. Providence, RI: American Mathematical
Society (AMS) (ISBN 978-0-8218-4699-5/pbk). xii, 234 p. $ 39.00 (2009).

Author profile:

Faddeev, Ludwig D.

Spellings: Faddeev, L.D. [141] Faddeev, L. [11] Faddeev, Ludwig D. [7] Faddeev, Ludwig [6] Faddeev, L. D. [6]
Author-Id: faddeev.ludwig-d
Publications: 171 including 36 Book(s) and 96 Journal Article(s)

MSC 2010

80	**81** · Quantum Theory
33	**35** · Partial differential equations (PDE)
31	**01** · History; biography
26	**37** · Dynamical systems and ergodic theory
25	**53** · Differential geometry

more ...

Journals

15	Russian Mathematical Surveys
6	Communications in Mathematical Physics
6	Journal of Soviet Mathematics
6	Theoretical and Mathematical Physics
5	Zapiski Nauchnykh Seminarov Leningradskogo Otdeleniya Matematicheskogo Instituta Imeni V. A. Steklova

more ...

Co-Authors

12	Takhtadzhyan, L.A.
8	Volkov, Alexandre Yu.
6	Osipov, Yu.S.
5	Gonchar, Andrei A.
5	Vershik, Anatoli M.

more ...

Publication Years

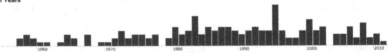

This very nice textbook contains the course developed and given by the first
author for students of mathematics at Leningrad University during the years
1968 through 1973. Thereafter this course was given by the second author.
The very clear formulated text introduces to quantum mechanics.

The first three sections are a straight and instructive summary of the canonical formalism of classical mechanics, especially underlining the multiplicative as well as Poissonian algebraic structures of the observables which become random variables when states are considered as probability measures
on phase space. The Liouville theorem for Hamiltonian flows is considered,
the uncertainty is introduced, and the dynamics of observables is contrasted
the Liouville equation. After consideration of the Planck and de Broglie relations, the double slit experiment and Heisenberg's heuristic introduction of

the uncertainty relation it is argued, that the commutativity of the multiplicative algebraic structure of observables cannot hold true for microsystems. In a finite dimensional matrix model the concepts of observables, spectral resolutions, the representation of Poisson brackets by commutators as well as states are introduced and the uncertainty relation is traced back to the commutator. The v. Neumann-Stone therem about irreducible representations of the Heisenberg commutation relations is cited which leads to the infinite dimensional Hilbert space $L^2(\mathbb{R}^3)$. The inductive part closes deriving the classical limit $\hbar \to 0$.

Author profile:

Yakubovskiĭ, O. A.

Spellings: Yakubovskiĭ, O. A. [1]
Author-Id: yakubovskii.o-a
Publications: 1 including 1 Book(s)

MSC 2010

1	81 · Quantum Theory

Co-Authors

1	Faddeev, Ludwig D.

Publication Years

2000		2010

The deductive part begins with one-dimensional systems: the motion of free wave packets, the harmonic oscillator, the motion in more general Potentials are treated. Concerning three dimensional Problems, symmetries, the rotational group, irreducible representations and spinor representations, and the hydrogen atom are considered. Perturbation theory, variational principles, and scattering theory including S-matrix theory follow. Further important feature of quantum theory like spectral representation w.r.t. a complete set of commuting observables, the spin of two electrons, the identity principle prepare the treatment of the helium atom and multi electron atoms. Mendeleev's periodic system and an appendix which considers the passage from Lagrangian to Hamiltonian mechanics conclude the textbook.

The authors have omitted to include complicated analytical details. They have given a very clear introduction to the main topics of quantum mechanics which can be best recommended to anyone, who wants to enter this field.

K.-E. Hellwig (Berlin)

Zbl 1200.22011
Ngô, Báo Châu
The fundamental lemma for Lie algebras.
Publ. Math., Inst. Hautes Étud. Sci. 111, 1-271 (2010).

It has been proved by Waldspurger that the fundamental lemma and transfer of orbital integrals for reductive groups over a non-archimedean local field can be deduced from the fundamental lemma for Lie algebras. Waldspurger also proved that validity of this fundamental lemma in the case of equal characteristics implies the validity in the case of unequal characteristics. The present article gives a proof in the case of equal characteristic p greater than twice the Coxeter number of the group.

The fundamental lemma has its origin in the problem of stabilisation of the trace formula for a reductive group over a global field, of characteristic p in the present case. It is an equality between a certain linear combination of local orbital integrals (a κ-orbital integral) for the adjoint representation of G and an orbital integral (a stable orbital integral) for an endoscopic group H of G.

The method is geometric and uses local and global arguments. Locally, one has the affine Springer fibers, introduced by Kazhdan and Lusztig. Stable orbital integrals and κ-orbital integrals can be expressed in terms of numbers of points in quotients of affine Springer fibers. Via Grothendieck's Lefschetz trace formula one obtains a cohomological interpretation of these integrals. As a global analogue of the affine Springer fibers, one has the fibers of the Hitchin fibration. An important tool here is the action on the space of the Hitchin fibration of a Picard stack over the base of the Hitchin fibration. A product formula gives the relation between the Hitchin fiber and the affine Springer fibers.

Geometric stabilisation is now formulated in terms of the perverse cohomology of the Hitchin fibration, or more precisely, of its restriction on a certain étale open subset (it is not the exact translation of the stabilisation of the trace formula).

The author proves a result on the support of the simple constituents of the perverse sheaves. The geometric stabilisation is first performed on a still smaller open subset, where the support theorem can be applied. Using this global result, the fundamental lemma is proved. Using the fundamental lemma, the geometric stabilisation is proved completely.

The author also proves a conjecture of Waldspurger ("the nonstandard lemma"), which, together with the fundamental lemma for Lie algebras, implies the twisted fundamental lemma.

The article is very well written.   *J. G. M. Mars (Utrecht)*

Author profile:

Ngô, Bao Châu

Spellings: Ngô, Bao Châu [10] Ngô, Báo Châu [4] Ngô, Bao-Châu [1] Ngô, Bao Chau [1] Ngô, Bao Châu [1]
Author-Id: ngo.bao-chau
Publications: 17 including 14 Journal Article(s)

MSC 2010

14	11 · Number theory
12	14 · Algebraic geometry
12	22 · Topological groups, Lie groups
2	20 · Group theory and generalizations

Journals

2	Annales Scientifiques de l'École Normale Supérieure. Quatrième Série
1	American Journal of Mathematics
1	Annales de l'Institut Fourier
1	Annals of Mathematics. Second Series
1	Compositio Mathematica

more ...

Co-Authors

2	Genestier, Alain
2	Haines, Thomas J.
1	Laumon, Gérard
1	Ngô, Tuân Dac
1	Polo, Patrick

Publication Years

1990   2000   2010